CONTENTS.

OFFICERS

OF THE

AMERICAN PHILOSOPHICAL SOCIETY,

FOR THE YEAR 1863.

PATRON,		His Excellency, the Governor of Pennsylvania.
PRESIDENT,		George B. Wood.
VICE-PRESIDENTS,		John C. Cresson, Isaac Lea, George Sharswood.
SECRETARIES,		Charles B. Trego, E. Otis Kendall, John L. Le Conte, J. Peter Lesley.
CURATORS,		Franklin Peale, Elias Durand, Joseph Carson.
TREASURER,		Charles B. Trego.
COUNSELLORS, elected for three years.	In 1861,	Alfred L. Elwyn, John Bell, Henry Coppée, Edward King.
	In 1862,	Isaac Hays, Robert E. Rogers, Henry C. Carey, Robert Bridges.
	In 1863,	Frederick Fraley, Robert Patterson, Daniel R. Goodwin, W. Parker Foulke.
LIBRARIAN,		J. Peter Lesley.

LIST OF MEMBERS

OF THE

AMERICAN PHILOSOPHICAL SOCIETY

ELECTED SINCE THE PUBLICATION OF THE ELEVENTH VOLUME.

Francis V. Hayden, M.D., of West Philadelphia.
Sidney George Fisher, of Philadelphia.
Sir Roderick Impey Murchison, F.R.S., of London.
Rev. Adam Sedgewick, F.R.S., of Cambridge, England.
Léonce Elie de Beaumont, of Paris.
Henry Milne Edwards, of Paris.
Dr. H. D. Bronn, of Heidelberg.
Dr. T. L. W. Bischoff, of Munich.
Dr. Hermann Von Meyer, of Frankfort on Maine.
Dr. Andreas Wagner, of Munich.
Dr. Joseph Hyrtl, of Vienna.
Sir William E. Logan, F.R.S., of Montreal.
Dr. Heinrich Rose, of Berlin.
Dr. George Jäger, of Stuttgardt.
Dr. St. Claire Deville, of Paris.
William Henry Harvey, M.D., F.R.S., of Dublin.
Dr. Jean Baptiste Dumas, of Paris.
Professor Edouard Verneuil, of Paris.
Dr. Claude Bernard, of Paris.
Daniel R. Goodwin, D.D., of Philadelphia.
Leo Lesquereux, of Columbus, Ohio.
John Lothrop Motley, LL.D., of Vienna.
Don Pascual de Gayangos, of Madrid.
John Curwen, M.D., of Harrisburg.
Dr. Charles Des Moulins, of Bordeaux.
Thomas Sterry Hunt, F.R.S., of Montreal, Canada.
Professor Paolo Volpicelli, of Rome.
Mirza Alexander Kasem Beg, of St. Petersburg.
Dr. Otto Böhtlingk, of St. Petersburg.
Dr. G. Forchhammer, of Copenhagen.
Dr. J. S. Steenstrup, of Copenhagen.
Dr. C. J. Thomsen, of Copenhagen.
Andrew Crombie Ramsay, F.R.S., of London.
Professor Edouard Desor, of Neuchâtel.
Dr. L. G. De Koninck, of Liège.
Dr. Joachim Barrande, of Prag.
Dr. Robert W. Bunsen, of Heidelberg.
Dr. William Hofmann, F.R.S., of London.
Dr. H. R. Göppert, of Breslau.
Dr. Alexander Braun, of Leipsig.
William John Hamilton, F.R.S., of London.
Sir William Jackson Hooker, F.R.S., of London.
Dr. J. J. Kaup, of Darmstadt.
Dr. J. Anthony Froude, of Oxford.
Hermann Lebert, M.D., of Breslau.
S. Weir Mitchell, M.D., of Philadelphia.
Dr. F. L. Otto Roehrig, of Philadelphia.
Lieut. Henry L. Abbot, Corps of Top. Eng. U. S. A.
Dr. Oswald Heer, of Zurich, Switzerland.
Dr. John Lindley, F.R.S., of London.
Dr. John Von Liebig, of Munich.
Dr. Frederick Wöhler, of Göttingen.
Professor J. D. Dawson, of Montreal, Canada.
Admiral Samuel F. Dupont, U. S. N.

Dr. George Engelmann, of St. Louis, Mo.
William S. Sullivant, Esq., of Columbus, Ohio.
Dr. Evan Pugh, Principal of Farmers' H. S. Pa.
Dr. Andrew A. Henderson, U. S. N.
Robert Cornelius, of Philadelphia.
Dr. Rudolph Virchow, of Berlin.
Dr. Frederic Theodore Frerichs, of Berlin.
Thomas Jefferson Lea, of Maryland.
Dr. Louis Stromeyer, of Hanover.
Dr. Karl Rokitansky, of Vienna.
Henry Winsor, Esq., of Philadelphia.
Dr. James Y. Simpson, of Edinburgh.
Dr. Théodore Schwann, of Liège.
Dr. Jacques L. Grimm, of Berlin.
Dr. Franz Bopp, of Berlin.
Dr. Ernest Renan, of Paris.
Dr. Max Müller, of Cambridge, England.
Joseph D. Whitney, S. G. California, San Francisco.
A. H. Worthen, State Geol. Illinois, Springfield.
Dr. Daniel Wilson, of Toronto, C. W.
Dr. Frederic Troyon, of Lausanne.
M. Boucher de Perthes, of Abbeville, France.
Prof. Pliny E. Chase, of Philadelphia.
Dr. I. I. Hayes, of West Philadelphia.
Dr. George Smith, of Delaware Co., Penna.
Hon. John M. Read, of Philadelphia.
Dr. Edward Jarvis, of Dorchester, Mass.

LIST OF MEMBERS

REPORTED DECEASED

SINCE THE PUBLICATION OF THE LAST VOLUME.

Hon. James K. Paulding.
Judge Thomas Sergeant, of Philadelphia.
Samuel D. Ingham, Esq.
Henry S. Tanner, Esq., of New York.
André-Marie-Constant Dumeril, of Paris.
Hon. Henry D. Gilpin, of Philadelphia.
Sig. Andrea Mustodixi, of Corfu.
Hartman Kuhn, Esq., of Philadelphia.
Major John Le Conte, U. S. A., of Philadelphia.
Steen Anderson De Billé, of Brussels.
Judge William Kent, of New York.
Dr. Samuel Moore, of Philadelphia.
Dr. William Harris, of Philadelphia.
Dr. Thomas Harris, of Philadelphia.
Dr. John W. Francis, of New York.
Don Pedro Cevallos, of Spain.
Prof. George Tucker, of Albemarle Co., Va.
Don Angel Calderon de la Barca, of Spain.
Joseph N. B. Von Abrahamson, of Copenhagen.
Sir John Forbes, F.R.S., of London.
F. Martinez de la Rosa, of Spain.
George M. Justice, of Philadelphia.
Sig. Hyacinth Carena, of Turin.
Charles J. Ingersoll, of Philadelphia.
George W. Bethune, D.D., of New York.
Dr. Karl C. Von Leonhard, of Heidelberg.
Dr. Edward Stanley, of London.
Duke Bernard Von Saxe-Weimar.
Dr. H. G. Bronn, of Heidelberg.
Hon. Samuel Breck, of Philadelphia.
Carl Adolf Agardh, of Lund, Sweden.
General O. M. Mitchell, U. S. A.
Edmé-François Jomard, of Paris.
Ellwood Morris, C. E., of North Carolina.
Col. Charles Ellet, U. S. V.
Capt. William R. Palmer, U. S. T. E.
Prof. James Renwick, of New York.
Col. J. J. Abert, U. S. Top. Engineers.
Dr. Carl Ludwig Rümker, of Hamburg.

EXTRACT

FROM THE LAWS OF THE SOCIETY RELATING TO THE TRANSACTIONS.

---

1. Every communication to the Society which may be considered as intended for a place in the Transactions, shall immediately be referred to a committee to consider and report thereon.

2. If the committee shall report in favor of publishing the communication, they shall make such corrections therein as they may judge necessary to fit it for the press; or, if they shall judge the publication of an abstract or extracts from the paper to be more eligible, they shall accompany their report with such abstract or extracts. But if the author do not approve of the corrections, abstract, or extracts reported by the committee, he shall be at liberty to withdraw his paper.

3. Communications not intended by their authors for publication in the Transactions, will be received by the Society, and the title or subject of them recorded; and, if they be in writing, they shall be filed by the secretaries.

4. The Transactions shall be published in numbers, at as short intervals as practicable, under the direction of the Committee of Publication, and in such a form as the Society shall from time to time direct; and every communication ordered to be published in the Transactions shall be immediately sent to the printer, and fifty copies thereof be given to the author as soon as printed.

5. The order in which papers are read shall determine their places in the Transactions, unless otherwise ordered by the Society; priority of date giving priority of location.

6. The expenses of publishing the Transactions shall be defrayed by subscriptions and sales, aided by such funds as the Society shall from time to time appropriate for that purpose.

COMMITTEE OF PUBLICATION.

ROBERT BRIDGES,
THOMAS P. JAMES,
EDWARD HARTSHORNE,
HENRY COPPEE,
CASPAR WISTAR.

ARTICLE IV.

INTELLECTUAL SYMBOLISM: A BASIS FOR SCIENCE.

BY PLINY EARLE CHASE, M.A.

Read, December 5th, 1862.

PREFACE.

"The nearer we come to Nature, the more does it seem to us that all our intellectual endowments are merely the echo of the Almighty Mind, and that the eternal archetypes of all manifestations of thought in man are found in the Creation of which he is the crowning work."—*Agassiz: Atlantic Monthly*, Vol. X, p. 94.

ויברא אלהים את האדם בצלמו בצלם אלהים ברא אתו׃

GENESIS 1 : 27.

THE intimate connection of religion and true philosophy is attested by the profound wisdom embodied in the teachings of the Bible, as well as by the most valuable records of all past history, yet there are many who unwisely try to divorce them, or to regard them as essentially antagonistic. But the tendency to inquiry is so natural, that any attempt to resist or suppress it will not only always prove futile, but it will even stimulate curiosity to an increased activity, which may be exerted in secret, and therefore, with greater danger of leading the investigator into pernicious error.

Those who discourage the discussion of religious or other dogmas, not only act in direct opposition to Paul's precept to the Thessalonians, "Prove all things; hold fast that which is good," but their action may be unwittingly instrumental in spreading the very evil that they wish to remove. Their opponents charge them at once with fearing investigation, and the plausibility of the charge often enlists the ardent sympathies of youthful inquirers, inducing them to listen to subtle reasonings that are cloaked in a fallacy too skilfully woven for them to unravel, and the fear of opprobrium deters them from discussing the arguments with those who might easily expose the fallacy.

True faith and true reason are handmaidens,—reason acknowledging its dependence on faith as the source of its authority, and faith demanding the assent of reason to no absurdities,—however important it may regard a belief in the mysterious and the incomprehensible. Both faith and reason may be often strengthened by the study of mysteries, and by forming such dim conceptions of their significance as may be traced in their faint shadowings, while both will be surely weakened if they become entirely self-reliant.

A fondness for philosophy is, then, fit cause for rejoicing, provided the spirit of inquiry is rightly guided. It is easy to show that no dictum of reason can be depended on as true, unless it can be traced to an infallible source, and that source can be none other than a Perfect Intelligence. All knowledge must, therefore, ultimately rest on revelation; the general knowledge of the race on a general revelation, and the special knowledge that may be adapted to newly arising needs of human liberty, on a special revelation. The question of faith, therefore, should not be, "Is this teaching perfectly comprehensible?" or "Is it such as unaided reason could have demonstrated for itself?" but "Is it such as the teacher knew to be true?"

Although all knowledge rests ultimately on direct revelation, it is modified and extended by faculties whose proper use may be regarded as a secondary or mediate revelation, and no satisfactory pursuit of knowledge is possible, without determining the extent, validity, and limitation of those faculties. On this account, Philosophy should begin with the study of Consciousness, and that study, like any other, may be most satisfactorily pursued, if it is pursued systematically. All system rests on laws of thought, and all laws imply relation. Relation may, therefore, be reasonably assumed as the natural basis of mental classification, and in the following pages I have attempted to show that a broad and comprehensive system may be developed from the necessary sequences of relation.

That the system as yet is far from complete, and that its present crude deductions will prove in many respects unsatisfactory, I am fully aware. One of the first teachings of faith is, as has been intimated, that the human Consciousness is not on the highest plane, but that the validity of its dicta rests on the infallibility of a higher Divine Consciousness, of which it is a faint and imperfect reflection. It is the instrument of philosophical investigation, but it can give us no information, except of that which is modified by its own relations. Parallel with it there may, perhaps, be always an accompanying Moral Inspiration, and an Active Genius, which bear to it relations similar to those of Motivity and Spontaneity to Rationality, but the province of these hypothetical parallel faculties we can never hope to investigate except through Analogy.

Generalization is always necessarily superficial, yet it often leads to the most satisfactory results. The very fact of its superficiality relieves memory from a heavy burden, and facilitates the use of symbols, by which the labors of reasoning and investigation are greatly abbreviated.

Whatever originality there may be in the symbolism by which I have endeavored to indicate the broadest of all possible generalizations, will naturally excite some degree of interest, but the final verdict as to its merits will rest on the answer to the old question,—*Cui bono?* To that question I dare not yet attempt to give or to seek a full response. If the system is, as I most fully believe, grounded on the eternal necessities of truth, it cannot be altogether insignificant, neither can it degenerate into a toy, fit only for the amusement of idle curiosity.

It may be said, indeed, that a mere notation can lead to no discoveries in Mental Science. This is in one sense true, but may not the same be said with equal, nay, with greater propriety, of the different notations of Mathematics, Logic, and Chemistry? The arbitrary signs of Algebra and the Calculus, do not contain in themselves the elements of any new truth, but they furnish a concise and precise language, by means of which, the results of a long process of investigation may be briefly expressed, and employed as the basis for new researches. A system of symbols that should render similar aid to Ethical, Social, and Intellectual Science, could hardly fail to yield important advantages.

A symbol in any case is not to be regarded as a box or wrapper in which some valuable but unknown truth is hidden, but it may be properly employed to represent, in the simplest possible form, an analysis that has already been made, and to keep the pure, unmixed results of that analysis so steadily in view, that they may be most conveniently used to facilitate farther investigation. Thus, although the Intellectual Symbols can convey very little meaning, until they are interpreted in familiar language, our operations with them may possibly lead us to reflect upon relations that had never before been observed, and by the study of these relations, new discoveries may be made.

If I am right in supposing the eternal necessity of an entity of the same order as Space and Time, for which I have proposed the name of Position,* the steps which led to its discovery, will well illustrate my meaning. Many have supposed "formal conditions of experience," or "logical antecedents of phenomena," different from Space and Time, but in every instance that has fallen under my notice, those conditions or antecedents have had either a subjective coloring, or a concrete reference. Take, for example, the ideas of personality, substance, cause, finitude, and figure. If there had never been manifestation, it is difficult to conceive either their necessary or their possible existence. But even if we suppose manifestation to be annihilated, Space, Time, and Position would still remain.

Since most, if not all our ideas of the Objective are derived from the material world, and our ideas of the Subjective from our own activity and its results, the necessary involution of space in the former, and of time in the latter, must have been evident to the earliest philosophical observers. But the necessity of a third form could never appear, until either the necessary triplicity of relativity, or the essential difference of relation from both the subjective and the objective, was perceived. The study of relativity led me to seek for the third formal condition, under which alone its triplicity was possible. Of my success in defining that condition by the sphere that I have assigned to Position, I must leave others to judge.

* See Chapter XII.

The following development would have been more entirely self-consistent, and for that reason it would, perhaps, have recommended itself to a more immediate and general approval, if it had stopped on the second plane below Consciousness, limiting its analysis to the determination of the three primary and nine secondary faculties. It would even then have covered a wider psychological field than has ever before been embraced in any systematic classification,—a field nearly as extensive as has ever been explored by any purely empirical philosophy. But I felt that it was desirable to indicate the direction of possible future discoveries, by a tentative analysis, that claims validity for none of its hypotheses, but seeks only to awaken an interest and stimulate an inquiry that may either verify or correct those hypotheses.

Still, if the success of this tentative analysis be judged by a comparison with previous attempts of a similar character, and not by a reference to the standard of absolute and necessary truth, it will, perhaps, be deemed sufficient to increase and justify the interest which may be first awakened by the mere novelty of the system. That the first Essay should fail of giving complete satisfaction in all its details is not strange, for the more minute the special subdivisions of any classification become, the slighter will be their distinguishing shades of difference, the more numerous and marked their various points of resemblance, and consequently, a greater degree of critical skill, and a more profound knowledge will be requisite, in order to make such an assignment of species as will stand every test of subsequent discovery. But if the fear of imperfection discourages us from using the resources at our command, we shall not only fall short of perfection ourselves, but we shall retard the progress of those who are to follow us, by neglecting to take the first steps which are necessary for all progress.

The spread of truth does not extinguish skepticism, but only banishes it to a remoter field. As the infinite limits the finite, so may the boundless realm of the doubtful and unknown be regarded as the limit of faith. The man who aims at a positive philosophy which shall embrace nothing that he does not understand,—refusing to accept anything on mere faith,—if he is consistent, will doubt everything, and even his boasted reason will be of no avail. But such consistency is fortunately impossible, for all are obliged practically to exercise a degree of faith which they often theoretically deny. Question closely and perseveringly as we may, we all finally arrive at simple truths that are accepted with implicit faith, not because our own authority is supreme, but because the clear perception of truth has been given us by the Supreme.

Any one who fully recognizes the relativity of Consciousness, and the correlations that it necessarily implies, will not only find absolute skepticism or atheism impossible, but by the very necessity that he discovers for revelation, he will be prepared to seek for the evidences of such revelation, and to consider favorably the arguments by which it is supported. If such a state of mind is more desirable than the carping, self-sufficient spirit, that scorns all external mental illumination, and tests the sunlight of eternal verities by the dim glimmer of its own flickering taper, I may reasonably ask for friendly criticisms on an attempt to determine some of the primary laws of relation, as well as for the co-operation of those who are interested in the development of those laws, and the determination of their consequences.

CHAPTER I.

DEFINITIONS AND FUNDAMENTAL RELATIONS.

1. SCIENCE is knowledge based on belief; FAITH is belief based upon revealed knowledge.*

2. Every object of human inquiry is an object either of Science or of Faith.

3. Belief is either absolute or relative.

4. Absolute belief is fundamental, simple, primary, necessary, independent, and irresistible. All absolute or self-evident beliefs are revelations from God,† being implanted in us by the Creator.

5. Relative belief is derivative, complex, subordinate, contingent, dependent, and debatable. Every relative belief is, however, the necessary resultant of all the data on which it is based, and all men would think alike *under the same circumstances*.‡

6. The sphere of knowledge is more extensive than that of absolute belief, embracing not only all primary, self-evident truths, but also every logical inference from absolute or indisputable premises.

7. Relative belief transcends knowledge, for it embraces inferences of every kind, whether logical or illogical, from premises true or false, together with all the convictions of feeling, and all the tenets of faith.

* Plato, *Republic*, B. 6, pp. 510–511, speaks of four different operations of the mind (ψυχή): intelligence (νόησις), demonstration (διάνοια), faith (πίστις), and conjecture (εἰκασία).

† Descartes appears to have been the first metaphysician who introduced into philosophy the evident truth, so beautifully expressed by Job, that men can acquire no knowledge except as "the inspiration of the Almighty giveth them understanding."

‡ Among the "circumstances" that determine erroneous belief, are improper assumption of premises, undetected fallacy, diseased mental action, ignorance or oversight of important facts, misunderstanding, hasty conclusion, defective generalization, wrong estimate of data. When men understand each other, they always agree in what they *know*, and they either harmonize in belief, or perceive that they would do so if they occupied the same standpoint.

8. Revelation is knowledge communicated in any manner by the Deity to his creatures. It is either direct or mediate.

9. Direct revelation is knowledge acquired without the aid of human reasoning, or the intervention of any human intelligence except our own.

10. Mediate revelation is knowledge which was originally acquired by direct revelation, and transmitted either by oral instruction, or by written record.

11. Belief, knowledge, faith, revelation, all imply thought and intelligence. The capacity of mind fixes the limits of knowledge.

12. All knowledge necessarily implies a dual existence, or an existence in two relations,* the existence of the knowing, and of the knowable.

13. From the dual or multiple, the mind naturally desires to ascend to the single or general, from the dependent variety to the independent, self-existent unity, which embraces and reconciles the plural or diverse.

14. That which knows, and that which is knowable, can only be united in a self-knowing intelligence. The highest conceivable unity is therefore a self-sufficient or "Absolute," self-conscious Being,—the Source or Originator of all actual as well as of all possible existence. Any other supposable highest intelligible or conceivable unity, must either be a unity of the knowing but unknowable, or of the knowable but unknowing, and therefore one of the forms of the highest duality, but by no means the all-embracing unity.†

15. Descartes, in his celebrated dictum, "Cogito Ergo Sum," was the first philosopher who clearly stated the fact, that consciousness necessarily involves the existence of the conscious being, and that all our knowledge must be based upon our personal consciousness. The same truth was more faintly shadowed forth in the "know thyself" of the Greek schools, but Descartes gave to the idea a clearness, simplicity, and fecundity of expression, that have revolutionized all metaphysical investigations.

* "If we appeal to consciousness, consciousness gives, even in the last analysis, in the unity of knowledge, a *duality of existence.*" *Hamilton, Discussions,* p. 66. "The necessary condition of intelligence is consciousness, that is, difference." *Cousin, Hist. of Mod. Phil.,* Vol. I, p. 88.

† We are apt to imagine, in the progress of philosophical investigations, that we discover a number of necessary but independent unities or realities, such as space, time, position, &c.; but a searching analysis will demonstrate that they are all merely forms of the knowable, standing in relation to our capacities of knowing, and that whatever necessity we may discover for their existence, is evidence of the necessary existence of a still higher unity.

If the essentiality and permanence of this duality, as well as its dependence on "the necessary condition of intelligence," is fully appreciated, Mahan's forcible statement of one of the strongest arguments in favor of immortality (p. 435), will seem almost axiomatic. "At death, not a particle of the physical organization, with which the soul is here connected, perishes. How unreasonable and absurd the supposition, that the soul, for which all else was made, is the only reality that then ceases to be."

16. All knowledge, therefore, starting from Intelligence, is limited by the nature and laws of Intelligence, and every Science must rest for its foundation on the Science of Mind. The Science of Science, which embraces all possible knowledge, was dignified by Socrates with the name of Philosophy, or the love of wisdom.

17. The simplest possible form of division is dual, but in treating of the faculties or capacities of Mind, there has been a very general recognition of triplicity. From the days of Pythagoras, who recognized in the soul three elements, Reason (νοῦς), Intelligence (φρήν), and Passion (θυμός),* to those of Hegel, who finds the manifestations of the *Idee* in Soul, Consciousness, and Reason, a fundamental ternary division has been adopted, with a marvellous unanimity which I can account for only by supposing it either to have been taught among the esoteric mysteries that shadow forth some of the earliest direct revelations to our race, or to have been founded on some obscure and dimly seen necessity of things.

18. For every general tendency, it is reasonable to suppose that there is some natural cause, yet no such cause appears to have been assigned or suspected, for the preference of any special form in the arrangement and classification of mental phenomena. There must be a great degree of uniformity in the facts that are made the objects of our study, and it is the duty of the critical investigator, to search for the law of which that uniformity is typical. "Facts are the words of God, and we may heap them together endlessly, but they will teach us little or nothing till we place them in their true relations, and recognize the thought that binds them together as a consistent whole."†

19. Among the many marvellous aphorisms of Aristotle, one of the most marvellous and productive is to be found in Book XI, Chap. XI, of his Metaphysics. "That which is changed is changed either from a characteristic‡ into a characteristic, or from a non-characteristic into a characteristic, or from a characteristic into a non-characteristic. I call that a characteristic which is made known by affirmation, so that it is necessary that

* Lewes. See also in *Anderson*, p. 76, the following citation from *Fragmenta Pythag. ex Theage in Opusculis Mythologicis.* "The soul consists of three parts: reason, irascible passion, and cupidity. Reason has subjected to it knowledge; passion, the bravery of strength; cupidity, appetite." Aristotle (ἠθικῶν Εὐδημίων, B. II, Chap. 7), says: "But of these three things, there would seem to be one; either according to longing (κατ' ὄρεξιν), or according to intention (κατὰ προαίρεσιν), or according to understanding (κατὰ διάνοιαν)." Many modern metaphysicians, adopting a more imperfect, because less comprehensive division, admit but three principal faculties of the mind: will, judgment, and understanding.

† *Agassiz: Atlantic Monthly,* July, 1862.

‡ I can think of no better translation for ὑποκείμενον than *characteristic* or *constituent.* The more obvious interpretations, *subject* and *substantial,* have been appropriated to denote more special meanings.

there should be three changes; for that from a non-characteristic into a non-characteristic is not a change."

20. On this view of the possible relations that can constitute the groundwork of propositions, the whole philosophy of Hegel appears to rest. He teaches "that everywhere the idea or notion appears first of all in its immediateness or intrinsic reality, that it then passes judgment upon itself, or becomes resolved into its opposite, and ultimately coalesces from out these antagonisms. From this very method results the whole structure or subdivision of the system. The Absolute, the being-thinking or *Idee*, has to pass through three momenta, and in the first place, to present itself as bare idea in and for itself. Secondly, in its differentiation or objective state, externality; and thirdly, as the idea that has returned from its externality into itself. In the first state, it is the purely logical *Idee*, the *thinking process* taken in the stricter sense as such in and for itself; in the second, it is the *Idee* in its externality, or departure from itself into a temporospatial disjunctivity, *i. e. nature;* and in the third, it is the *mind* or intelligence. Accordingly the whole of philosophy, or the thinking process, which has comprehended itself in this its active state, has three cardinal divisions, the Logic, which with Hegel, as is readily seen, implies also *Metaphysics;* the Philosophy of Nature, and Philosophy of Mind. . . .

21. "Within each of these three cardinal divisions, the same rhythmical movement repeats itself, and produces a like threefold division. The Logic has to deal (*a*) with the first immediateness, or with being; (*b*) this divides itself into the antagonism of essence and existence, and these finally coalesce together to form the idea (*Begriff*), with which we have already become acquainted, both in its real as well as ideal import, as the living circulation of momenta including itself within itself."*

22. These expositions of some of the most profound thoughts of Greek and German philosophy are well worthy of attention, and notwithstanding the obscurity with which they are clothed by foreign idiomatic forms of thinking and expression, it is easy to discern, in the general idea of which they are special and profitable applications, the grand, fundamental idea of all philosophy,—the idea of relativity, as the basis of analysis and synthesis. This idea, in its broadest generality, may be thus stated with mathematical vigor.

23. Given a and b, there can be four, and only four relations of antecedent and consequent, viz., aa, ab, ba, bb.

Of these four possible relations, only three can concern either of the given terms, *e. g.*, a is involved only in the three relations, aa, ab, ba, and b only in ab, ba, bb.

24. The fundamental relations involved in Philosophy are, as we have already seen, the

* Chalybäus, pp. 343–5.

relations of the knowing, or as it is technically called, the *Subjective*, and of the knowable, or the *Objective.**

25. These relations may be thus designated:

1. The Subjective-Subjective, in which the Subjective is both antecedent and consequent.

2. The Subjective-Objective, in which the Subjective is antecedent, and the Objective is consequent.

3. The Objective-Subjective, in which the Objective is antecedent, and the Subjective is consequent.

4. The Objective-Objective, in which the Objective is both antecedent and consequent.

26. In all inquiries connected with the science of Mind, the subjective is necessarily involved, and in consequence of this necessity, mental investigations can be in no way concerned with the last of these four relations. Of the merely objective-objective, we can have no possible knowledge and no positive conception, all our ideas of the action of objects upon each other, or of their mutual relations, being derived from the union or comparison of objective-subjective and subjective-objective impressions.

27. There are then left for our consideration, but three primitive relations, each of which represents a distinct phase or form of the Subjective Mind.

28. The *essential* attribute of Mind is CONSCIOUSNESS.†

29. There may be forms of immaterial substance that are devoid of Consciousness, of which Force is perhaps one, but we give the name of Mind only to that portion of our being which has the power of perceiving its own operations, and the impressions that are made upon it. We can neither feel, act, nor think, without being conscious at the moment, of the feeling, action, or thought. It is true that the conscious impression is often faint and momentary, and that it often slips instantly from our memory unless there is

* "In the philosophy of mind, *subjective* denotes what is to be referred to the thinking subject, the Ego; *objective* what belongs to the object of thought, the Non Ego. . . . The exact distinction of *subject* and *object* was first made by the schoolmen; and to the schoolmen the vulgar languages are principally indebted for what precision and analytic subtilty they possess." *Hamilton: Discussions*, p. 13; see also *Cousin: El. of Psychology*, p. 358.

The subjective can become objective to itself, but the objective cannot become subjective. The subjective or intelligent is therefore supreme.

† "The fact of consciousness is a complex phenomenon, composed of three terms: the *me* and the *not-me*, bounded, limited, finite; again, the idea of the infinite; and still again, the idea of the relation of the *me* and the *not-me*, that is, of the finite to the infinite." *Cousin: Hist. of Mod. Philos.*, Vol. I, p. 126.

"The first fact with regard to the soul is that it is intelligent and vocal,—that it is not merely a subject, but also an organ of THAT WHICH KNOWS in the universe." *D. A. Wasson: Atlantic Monthly*, Vol. XI, p. 40.

something to fix the attention, but we can study mind only in Consciousness, and it is entirely out of our power to form any notion of the nature or attributes of unconscious mind.

30. Consciousness, in its action, involves duration, or Time. Every conscious process has a beginning, an advance, and an end. The relations of the Subjective, are therefore relations in time,—the origin of the relation determining the chronological antecedent,—and the termination of the relation, the chronological consequent.

31. In the Objective-Subjective relation, the impulse commencing externally and terminating in Consciousness, our attention is aroused, and we are induced to exercise our activity in various ways. To this form of Consciousness, which corresponds very nearly to the Passion (θυμός) of Pythagoras, the name of Passivity or Receptivity might be given, to designate the condition of the mind as the recipient of an impulse not originating in itself. But as the simplest exercise of Consciousness involves some degree of activity, and as the aroused attention tends to incite increased activity, the term Motivity seems more appropriate.

32. In the Subjective-Subjective relation, the impulse begins and ends within Consciousness, which is said to act "of its own accord," or "spontaneously." I propose to designate this form of the subjective by the term Spontaneity.

33. In the Subjective-Objective relation, we are subjectively conscious of an effort commencing in our own minds, but tending towards the objective, an effort to perceive, know, understand, the nature of the object, or the proper mode of using it to accomplish some particular end that we have in view. This is especially an Intellectual or Rational effort, and the term Rationality seems peculiarly fit for the form of Consciousness in which this effort originates.*

34. It is extremely difficult, if not absolutely impossible, to bring this primary division of Consciousness purely under our observation, because we can never observe the mind when it is merely motive, spontaneous, or rational. The very effort of observing involves a subjective exercise of Spontaneity and Rationality, and renders the mind while using its subjective energies, the object of its own observation. The effort to penetrate this labyrinth of complicated objective and subjective influences is perplexing, but no more so than

* I do not remember to have seen the boundaries of the primary divisions of Consciousness more clearly indicated, than by Mahan (p. 15), who employs the terms, "Intellect or Intelligence, Sensibility or Sensitivity, and Will. To the Intellect we refer all the phenomena of *thought*, of every kind, degree, and modification. To the Sensibility we refer all *feelings*, such as sensations, emotions, desires, and affections. To the Will we refer all mental *determinations*, such as volitions, choices, purposes, &c." Although this division, which is based upon pure observation, does not precisely correspond with our own, the resemblance is sufficiently striking to afford a very satisfactory confirmation of our theory.

the study of the relation that exists between our subjective ideas of matter and its attributes, and the objective attributes of matter as they exist in themselves, and if our clue is not long enough to thread the entire maze, it may at least enable us to effect an entrance, and gradually to explore a goodly portion of the labyrinth.

35. Though Motivity, Spontaneity, and Rationality may never be seen in pure and separate activity, in their combined action we can always, and usually without much difficulty, recognize one of the three as predominant. The respective degrees of influence severally exerted by the three Conscious forms, furnish us with a basis for division into primary faculties, and for subdivision to any required extent, according to subjective or objective tendencies, or rather according to motive, spontaneous, or rational resemblances. If any difficulty in precisely limiting and defining any particular faculty should appear discouraging, it may be well to glance at the various attempts that have been made at classification in the Natural Sciences, and to the proverbial difficulties that surround every attempt at system, introducing perturbations into most of our calculations, and producing exceptions to all general rules. Shall we discard the division of Physical Nature into three kingdoms, because it is impossible to determine the point at which the mineral is clothed with vegetable life, or to mark the precise boundary between the zoophyte and the plant? If Cuvier and Agassiz differ in opinion as to the genus or species to which a particular animal should be assigned, shall we pronounce the Science of Natural History worthless?

36. Our primary division of Consciousness has been logically deduced from a consideration of the relations which it necessarily assumes to the objective, but these relations do not in any way change the essential nature of the related terms. Like Consciousness itself, each of its subdivisions is subjective, and may be analyzed in its turn by regarding the modifications it assumes under different relations, as determining or determined by the objective, or as acting under subjective influences for purely subjective ends. If, then, we designate the Subjective under the objective-subjective relation by M, under the subjective-subjective by S, and under the subjective-objective by R, Motivity, Spontaneity, and Rationality may be severally indicated by the simple symbols, M, S, R, and their immediate subdivisions by MM, MS, MR,—SM, SS, SR,—RM, RS, RR. Extending this plan of subdivision, we obtain the following symbolic schema, which may be continued indefinitely, marking a precise, well-defined, and philosophical arrangement of the mental faculties.

37. Consciousness.

Motivity (M).			Spontaneity (S).			Rationality (R).		
MM	MS	MR	SM	SS	SR	RM	RS	RR
MMM	MSM	MRM	SMM	SSM	SRM	RMM	RSM	RRM
MMS	MSS	MRS	SMS	SSS	SRS	RMS	RSS	RRS
MMR	MSR	MRR	SMR	SSR	SRR	RMR	RSR	RRR

38. If we assign names to each of these symbols indicative of their exact significance, we may make an exhaustive catalogue of the powers of the mind. In selecting those names, it will be well to appropriate as far as possible, those that are already in use; for new and unfamiliar names would have no advantage over the simple symbols, and they would cumber the memory without conveying so distinct ideas as the symbols that they were supposed to illustrate and explain.

39. But when terms have become familiar, and have acquired a meaning somewhat precise, it is not wise to discard them altogether, especially if we can submit them to a system that will render their significance still more definite.* There are equally valid reasons for allowing observation and experiment to precede theory and scientific classification, in mental as in physical science. Every diligent observer will discover interesting and valuable facts, in whatever direction his inquiries may be turned, and the researches of mental philosophers will be found to have developed a mass of information and nomenclature, much more varied than that aggregate of physical ideas on which the modern natural philosophy was built. If this information were all arranged in accordance with any system, however imperfect, so that it could be readily learned, it would afford great aid to those who desire to pursue metaphysical investigations, and its value for educational purposes would be inestimable.†

40. Dr. Reid, starting from the division of the mental faculties into those of *understanding* and those of *will*,—a division which Hamilton traces "to the classification taken in the Aristotelic school, of the powers into *gnostic* or cognitive, and *orectic*, or appetent," recognizes the mutuality of the faculties in the following remarks:

41. "As the mind exerts some degree of activity even in the operations of understanding, so it is certain that there can be no act of will, which is not accompanied by some act of understanding. The will must have an object, and that object must be apprehended

* "The common terms of a language represent the results attained by the experience of all preceding ages."

† "But that each one says something concerning nature; and though each singly adds nothing or little to it, yet from all collected, there is some magnitude." *Aristotle: Metaphysics*, B. 2, chap. 1.

or conceived in the understanding. It is therefore to be remarked, that in most if not all operations of the mind, both faculties concur, and we range the operation under that faculty which has the largest share in it."*

42. An arrangement of faculties based upon this view, would appear to be quantitative, and the order of the symbols would indicate the supposed relative degrees of influence, exerted by Motivity, Spontaneity, and Rationality in mutual action. Nor would the results of such a hypothesis be altogether unsatisfactory, for in every mental operation, we could trace predominant traits, and secondary and subordinate characteristics, sufficiently marked to enable us to assign symbols in such order, as would fix its position in the schema we have adopted, and thus indicate the meaning that we attached to the name by which we described the operation. But according to our explanation of the symbols, the first letter indicates a modification of simple Consciousness, and the subsequent letters, *analogous* (not *identical*), modifications of the subordinate forms of Consciousness. By the former hypothesis, the symbol RM (Rationality-motive), would indicate a faculty in which Rationality was principally concerned, and Motivity in a smaller degree; by the latter, it would denote that form of Rationality which is modified by relation, in a manner similar to the modification of Consciousness in Motivity. The difference may be slight, and in the present state of Mental Science, perhaps inappreciable, still there is a difference.

* To this passage, Sir William Hamilton appends the following note: "It should be always remembered that the various mental energies are all only possible in and through each other" [should we not rather say, in and through Consciousness?] "and that our psychological analyses do not suppose any real distinction of the operations which we discriminate by different names. Thought and volition can no more be exerted apart, than the sides and angles of a square can exist separately from each other." *Reid*, p. 242. This fundamental characteristic of mental manifestation facilitates our analysis, by rendering a system that would otherwise appear arbitrary and artificial, perfectly philosophical and natural.

We might suppose, for instance, a classification of physical phenomena, based on the three dimensions, length, breadth, and thickness, that should represent all the facts of natural philosophy by combinations and permutations of the symbols L, B, T. Or we might undertake to explain the functions of civil government, by similarly combining the symbols M, A, D, which would severally represent monarchy, aristocracy, and democracy. But would those symbols, in either instance, denote *necessary* relations, the *only* necessary relations, and relations that are necessarily repeated and continued at each successive step of subdivision? We have seen that all this is true of the symbols M, S, R. Relativity is essential to Consciousness, and in whatever way we suppose Consciousness to be modified, it is Consciousness still, with a capacity for action under three and only three general relations.

CHAPTER II.

PRIMARY FACULTIES.

43. Although metaphysical writers have generally turned their attention almost exclusively to the rational phenomena, they have often recognized both the motive and the spontaneous element of Consciousness. The following are some of the prominent terms that philosophers have employed, to designate mental states that they have specially observed, with the symbol attached to each that seems most precisely to indicate its meaning:

Propensity (MM),	Instinct (SM),	Perception (RM),
Desire (MS),	Will (SS),	Judgment (RS),
Sentiment (MR),	Energy (SR),	Understanding (RR).

44. In order to determine the correctness of this relative assignment, it may be well to examine each of the terms somewhat carefully.

45. Propensity, as defined by Webster, denotes "bent of mind, natural or acquired; inclination; natural tendency."

46. Comte, treating of the phrenological subdivision of the affective faculties into propensities and sentiments, says that "the first and fundamental class" [propensities] "relates to the individual alone, or at most, to the family, regarded successively in its principal needs of preservation, such as reproduction, the rearing of the young, the mode of alimentation, of habitation," &c.*

47. Combe says, "All the propensities are blind," and "the faculties of the propensities and sentiments cannot be excited to activity directly by a mere act of the will," but "each faculty may be roused into activity by the *presentment of its appropriate objects*."†

48. Whatever we may think of the comparative accuracy of these several definitions, there can be little doubt that their authors regarded Propensity as directly subject to an external, objective stimulus, and it may, therefore, be ranked unhesitatingly under Motivity. Inasmuch as it denotes a mere tendency, without any perceptible (quantitative) element of Spontaneity or Rationality, it may well be regarded as the simplest or motive form of Motivity (Motivity affected), the symbol of which is MM.

49. Desire, "even when its object is some action of our own; is only an incitement to will, but it is not volition."‡ (Motivity, but not Spontaneity, though somewhat like it.)

* *Positive Philosophy*, pp. 389–390.

† *Lect. on Phrenology*, pp. 140, 277, 278.

‡ Reid, p. 532.

50. "The uneasiness a man finds in himself upon the absence of anything, whose present enjoyment carries the sense of delight with it, is that we call desire."*

51. "That which immediately determines the will from time to time, to every voluntary action, is the uneasiness of desire fixed on some absent good."*

52. The symbol MS accords well with these definitions, denoting Motivity with a special tendency to voluntary or spontaneous action.

53. SENTIMENT "supposes the existence of some social relations, either among individuals of a different species, or especially between individuals of the same species apart from sex, and determines the character which the tendencies of the animal must impress on each of these relations, whether transient or permanent."†

54. "Authors who place moral approbation in feeling only, very often use the word *Sentiment*, to express feeling without judgment. This I take likewise to be an abuse of a word. Our moral determinations may, with propriety, be called *moral sentiments*; for the word *sentiment*, in the English language, never as I conceive, signifies mere feeling, but *judgment accompanied* with feeling."‡ (Say rather, feeling implying or suggesting the idea of judgment.)

55. These definitions justify us in regarding Sentiment as a feeling or affection of Consciousness, excited by any appropriate object, and tending to produce action in accordance with our position as social and rational beings. It therefore represents Motivity, tending towards subjective action with a rational object or end, and its appropriate symbol is MR.

56. INSTINCT, as defined by Reid, is "a natural blind impulse to certain actions, without having any end in view, without deliberation, and very often without any conception of what we do." This is exemplified in "that natural instinct by which a man who has lost his balance and begins to fall, makes a sudden effort to recover himself, without any intention or deliberation."§

57. Hamilton says, "An Instinct is an agent which proposes blindly and ignorantly, a work of intelligence and knowledge."||

58. Comte, in noticing the relation of intelligence to instinct, observes that "the only meaning that can be attached to the word *instinct*, is any spontaneous impulse¶ in a determinate direction, independently of any foreign influence. In this primitive sense, the

* Locke, v. 1, pp. 149 and 160.

† Comte, p. 390.

‡ Reid, p. 674. On this paragraph, Sir William Hamilton remarks: "This is too unqualified an assertion. The term *Sentiment* is in English applied to the *higher feelings*."

§ Reid, p. 568.

|| Reid, p. 761.

¶ Spontaneous impulse = Spontaneity-Motive, SM.

term evidently applies to the proper and direct activity of any faculty whatever, intellectual as well as affective, and it therefore does not conflict with the term *intelligence* in any way, as we so often see when we speak of those who without any education, manifest a marked talent for music, painting, mathematics, &c. In this way there is instinct, or rather there are instincts in man as much or more than in brutes."*

59. We may infer from the foregoing definitions, that the prominent or general characteristic of Instinct, is a tendency to spontaneous action, while the secondary or specific characteristic, is "a natural blind impulse," analogous to Propensity. If this impulse was considered as the most obvious feature of Instinct, its symbol would be MS, but if, as I believe, the first idea suggested by the term, is that of some kind of active potentiality, it represents the motive form of Spontaneity, and its symbol is SM. It will be found both interesting and useful, to observe the analogy and the quantitative distinction between Desire and Instinct,—the spontaneous-motivity and the motive-spontaneity,—the elements of each being the same, but Motivity being more prominent in the former, and Spontaneity in the latter.

60. WILL is so purely subjective, that its place may be assigned without hesitation under Spontaneity, of which it may be regarded as the spontaneous form, and its symbol is therefore SS.

61. It is difficult by any definition, to describe a faculty that is so familiar to every one by its constant action, so as to give any clearer idea of its limits, than we obtain by the very position we have given it in our schema, as the subjective, absolute, or spontaneous form of the subjective-subjective.

62. Bacon says: "The knowledge which respecteth the faculties of the mind of man, is of two kinds; the one respecting his understanding and reason, and the other his will, appetite, and affection; whereof the former produceth direction or desire, the latter action or execution."†

63. Locke says: "This at least I think evident, that we find in ourselves a power to begin or forbear, continue or end several actions of our mind and emotions of our bodies, barely by a thought or preference of the mind ordering, or as it were, commanding the doing or not doing such or such a particular action. This power which the mind has thus to order the consideration of any idea, or the forbearing to consider it, or to prefer the emotion of any part of the body to its rest, and *vice versâ*, in any particular instance, is that which we call the will."‡

64. According to Reid, "Every man is conscious of a power to determine, in things

* Comte, pp. 585–6. Taylor, *Elements of Thought*, p. 105, thinks that Instinct "cannot be imagined to reside in the animal."

† Bacon's Works, Vol. I, p. 206.

‡ Locke, B. 2, c. 21, § 5.

which he conceives to depend upon his determination. To this power we give the name of *Will*."*

65. Cousin observes as follows: "The peculiar characteristic of the me is causality, or will, since we refer to ourselves, we impute to ourselves only what we cause, and we cause only what we will. To will, to cause, to exist for ourselves,—these are synonymous expressions for the same fact, which comprises at once will, causality, and personality. . . . . The phenomenon of will presents the following elements, 1, to decide upon an act to be performed; 2, to deliberate; 3, to resolve. Now if we look at it, it is reason which composes the first element entirely, and even the second; for it is reason also which deliberates, but it is not reason which resolves and determines."†

66. Jouffroy says: "To *direct* and *to correct*, such is then the double action of the personal power over the development of our faculties. . . . *The personal faculty* (or that supreme power that we have to make use of ourselves, and of the capacities which are in us, and to dispose of them), is known under the names of *liberty* and of will, which designate it but imperfectly."‡

67. ENERGY, if we look merely to its etymological derivation, would appear to imply activity. Sir William Hamilton says: "Energy is often ignorantly used in English for force. . . *Operation*, *Act*, *Energy*, are nearly convertible terms, and are opposed to *Faculty*, as the *actual* to the *potential*."§ If this position of the distinguished philosopher is impregnable, there would be a manifest impropriety in ranking Energy among the faculties of the mind.

68. But whatever may have been the original meaning of the word, it is evident that its ordinary acceptation at the present day, does not necessarily involve the idea of activity. Moreover, by a common metonymy, the same word is often used to denote a faculty, or power of mental action, and also to designate the specific act of that faculty. Thus desire, sentiment, instinct, will, perception, judgment, are all employed with a twofold meaning,—one subjective, and the other objective; and if there is no impropriety in applying the term judgment indiscriminately, to the faculty of judging and to the decision of that faculty, there can be none in using the term Energy to denote the faculty of acting for a fixed purpose, as well as to denote the action itself.

69. Energy is variously defined by our principal lexicographers, as "power, inherent or exerted;" "force;" "vigor;" "operation;" "strength;" "efficacy;" "faculty." An energetic man is equally energetic, whether he is active or at rest; he is one who has the faculty of intelligent and successful activity, which he may exercise, like his other faculties, at his own pleasure.

* Reid, p. 530.

† Cousin, pp. 384, 386.

‡ *Melanges Philosophiques*, 2de edit., pp. 345–349.

§ Reid, pp. 515, 221.

70. Energy, therefore, in ordinary philosophical diction, as well as by common usage, seems to involve,

1. The power of subjective activity, or Spontaneity;

2. The power of directing subjective activity to a special or rational end. It is, then, the faculty of Spontaneity-rational, and its symbol is SR.

71. Rationality being, as we have already stated, specially concerned with the acquisition of knowledge, its subordinate faculties should be adapted to every possible method of acquisition.

72. Now Rationality may attain its end in three, and only three ways, viz.:

1. By the acquisition of new facts,—the rational Me affected by the Not Me.

2. By combining or comparing two or more facts, drawn from the storehouse of Consciousness, in order to discover new forms of truth from their relations, the Rational Me acting within or upon itself.

3. By the examination of facts or conclusions, for the intelligent determination of their full objective meaning,—the Rational Me overstepping the bounds of experience, to declare the reality of the Not Me.

73. In the acquisition of new facts, the rational Consciousness is sub-passive and receptive, influenced by the objective, and simply percipient of the fact which is presented for its cognizance. This motive form of Rationality is usually called Perception.

74. In comparing the facts which, either by original constitution or by appropriation, have become a portion of its own treasury of knowledge, the rational Consciousness is specially subjective, its action originating and terminating within its own borders. This spontaneous form of Rationality has the same characteristics as the faculty of Judgment.

75. In ascertaining objective significance, the subjective Rationality assumes an objective tendency, and is evidently in its *affecting* or rational form. This form corresponds to the faculty of Understanding.

76. Perception is confounded by Locke with thinking, and with the act of the Understanding.* He remarks, however, that "thinking, in the propriety of the English tongue, signifies that sort of operation in the mind about its ideas, wherein the mind is active; when it, with some degree of voluntary attention, considers anything. For in bare, naked perception, the mind is for the most part, only passive, and what it perceives, it cannot avoid perceiving."†

77. Kant defines perception as "empirical consciousness," and he remarks that phenomena, as objects of perception, "contain in themselves, besides the intuition, also matter for an object in general (whereby something existing in space or time is represented)."‡

* Essay, Vol. I, pp. 90, 98, 152. † Ibid. p. 98. ‡ Hayward's translation, p. 138.

78. Reid says perception "hath always an object distinct from the act by which it is perceived; an object which may exist, whether it be perceived or not."*

79. Combe regards perception as the act of a faculty which recognizes an object on presentation,—" the lowest degree of activity of the intellectual faculties."†

80. Hamilton says: "*External Perception* or *Perception* simply, is the faculty *presentative* or *intuitive* of the phenomena of the Non-Ego or Matter,—if there be any *intuitive* apprehension allowed of the Non-Ego at all. *Internal Perception* or *Self-Consciousness* is the faculty *presentative* or *intuitive* of the phenomena of the Ego or Mind."‡

81. We infer, therefore, that philosophers unite in regarding perception,

1. As a subjective-objective or rational faculty.

2. As sub-passive, empirical, active in the lowest degree, tending to incite rationality to greater activity,—attributes which should characterize Rationality-motive. We are therefore confirmed in our previous assignment of its position, as the representative of the symbol RM.

82. Judgment is undoubtedly a rational faculty, involving a special active exercise of our subjective powers, such as should belong to the absolute or spontaneous form of Rationality. Its symbol is therefore RS. Compare this localization with the following definitions.

83. "The faculty which God has given man to supply the want of clear and certain knowledge, in cases where that cannot be had, is judgment; whereby the mind takes its ideas to agree or disagree, or which is the same, any proposition to be true or false, without perceiving a demonstrative evidence in the proof.§

84. "Judgment is the thinking or taking two ideas to agree or disagree, by the intervention of one or more ideas, whose certain agreement or disagreement with them it does not perceive, but hath observed to be frequent and usual."||

85. "*Judgment* is, therefore, the mediate cognition of an object, consequently the representation of a representation of it. In every judgment there is a conception, which is valid for many, and under such many comprehends also a given representation, which last thing then is referred immediately to the object. . . . But we can reduce all actions of the Understanding to judgments, so that the *Understanding* in general can be represented as a *faculty of judging*."¶

86. "The definition commonly given of judgment, by the more ancient writers in logic

* Reid, p. 183.

† Op. citat., p. 284.

‡ Reid, Note B, § I, 8, p. 809.

§ Locke, v. 2, p. 427.

|| Locke, v. 2, p. 445.

¶ Kant, p. 61.

was, that it is *an act of the mind whereby one thing is affirmed or denied of another.* I believe this is as good a definition of it as can be given."*

87. "In treatises of logic, judgment is commonly defined to be an act of the mind, by which one thing is affirmed or denied of another; a definition which, though not unexceptionable, is perhaps less so than most that have been given on similar occasions."†

88. "The arts intellectual are four in number, divided according to the ends whereunto they are referred; for man's labor is to invent that which is sought or propounded; or to judge that which is invented; or to retain that which is judged; or to deliver over that which is retained. So as the arts must be four; art of inquiry or invention; art of examination or judgment; art of custody or memory; and art of elocution, or tradition."‡

89. "Judgment is the action of the mind in deciding or pronouncing, concerning two things, when placed in comparison, that they are equal or unequal, like or unlike; that the one contains the other, or bears such or such a relation to it. It is by *successive judgments*, or by the regular comparing of one idea or notion with another, until we reach some one which at first was seen in the distance, that a process of reasoning is carried on."§

90. "Sound judgment is feeling rightly and perceiving correctly. The reflective faculties are the judges, but they depend on the other faculties for correct data."||

91. This faculty of Judgment corresponds in part if not wholly with what Hamilton calls the Elaborative Faculty. He says: "These four acts of acquisition, conservation, reproduction, and representation, form a class of faculties which we may call the Subsidiary, as furnishing the materials to a higher faculty, the function of which is to elaborate these materials. This elaborative or discursive faculty is Comparison; for under Comparison may be comprised all the acts of Synthesis and Analysis, Generalization and Abstraction, Judgment and Reasoning. Comparison, or the Elaborative or Discursive Faculty, corresponds to the Διάνοια of the Greeks, to the *Verstand* of the Germans. This faculty is Thought Proper; and Logic, as we shall see, is the science conversant about its laws."¶

92. The difference between Energy (SR), and Judgment (RS), according to our schema, is that in the former the Spontaneous element, and in the latter the Rational, is the more prominent; the one being the faculty of subjective activity for a rational purpose or end, the other the faculty of rational action for subjective improvement or gratification.

93. UNDERSTANDING is the rational form of Rationality,—the supreme faculty of Intelligence to which Perception and Judgment are both subservient. Its symbol is RR.

94. "The commandment of knowledge is yet higher than the commandment over the

* Reid, p. 413.
† Stewart, p. 349.
‡ Bacon, Vol. I, p. 207.
§ Taylor, *El. of Thought*, p. 110.
|| Combe, p. 290.
¶ Hamilton: *Metaphysics*, p. 284.

will, for it is a commandment over the reason, belief, and understanding of man, which is the highest part of the mind, and giveth law to the will itself."*

95. "As the fancy is the apprehension or seizing of an object, the reason a combination or distinction, so the understanding is the faculty which penetrates, and in its highest degree, clearly sees through its object. We understand a phenomenon, a sensation, an object, when we have discovered its inmost meaning, its peculiar character and proper significance. And the same is the case, even when this object be a speech and communication addressed to us,—a word or discourse given us to extract its meaning. If we have discerned the design which is involved in such a communication, its real meaning and purpose, then may we be said to have understood it, even though some minutiæ in the expression may still remain unintelligible, which, as not belonging essentially to the whole, we put aside and leave unconsidered."†

96. "But there is a spirit in man; and the inspiration of the Almighty giveth them understanding."‡

97. In Understanding, as in Propensity, the objective element largely predominates; the former being almost exclusively objective in its tendency,—the latter in its origin. Accordingly the mind is more nearly passive in the exercise of this faculty, than in either Perception or Judgment, for our activity proceeds so far, and so far only as the subjective is involved.§

98. As Motivity and Spontaneity are both developed before Rationality,—the animal and physical operating as conditions for intellectual growth,—so this sovereign faculty of Rationality matures more slowly, and attains its highest development at a later period of life than any of the other primary faculties. Indeed, in the mass of mankind, the Understanding always remains feeble, and, as it were infantile,—a power prophetic in its latent capabilities, of a higher state of existence, in which it will be called into full and proper exercise.

99. The powers of Perception and Judgment, are to some extent employed at almost every instant of our lives,—so frequently, that our familiarity with their objects is often mistaken for a complete understanding. How serious this mistake, let the meagreness of science, and the vast field of tempting speculation that continually allures baffled speculators, testify. Of how many men may it be said at all times, and how often may it

* Bacon, Vol. I, p. 182.

† Schlegel, p. 54.

‡ Job, c. 32, v. 8.

§ Mahan (pp. 214–18), adopting the views of Coleridge, says: "The Understanding is the faculty of *believing*. The Reason is the faculty of *knowing*." This definition would not justify the use of the symbol RR, but it shows the need of some system for precisely indicating the meaning of the terms that we employ.

be said of all men, that they "seeing see not, and hearing they hear not, neither do they understand."

100. The relation between the objective and subjective, involved in the convictions of the Understanding, is and will perhaps always remain wholly incomprehensible. The attempt to explain it by "ideas," "sensible impressions," or the intervention of imaginary objects which are neither material nor immaterial, serves only to increase the obscurity, which envelops the mode of mutual action between mind and matter. We can only say that those convictions are irresistible, that they are in accordance with the intellectual nature given to us by the Creator, and as such we ground our faith upon them, as revelations from that Supreme Intelligence, "for whom are all things, and by whom are all things." "The entrance of thy words giveth light; it giveth understanding to the simple."*

101. Of the Divine Reason, that gives to the Understanding all its convictions,—the light "which lighteth every man that cometh into the world," Cousin discourses beautifully (though somewhat questionably), as follows:

102. "When we come to interrogate reason about itself, to inquire into its own principle and the source of that absolute authority which characterizes it, we are forced to recognize that this reason is not ours, not constituted by us. It is not in our power, it is not in the power of our will to cause the reason to give us such or such a truth, or not to give us them. Independent of our will, reason intervenes, and when certain conditions are fulfilled, gives us, I might say, imposes upon us these truths. The reason makes its appearance in us, though it is not ourselves, and in no way can it be confounded with our personality. Reason is impersonal. Whence then comes this wonderful guest within us, and what is the principle of this reason which enlightens us, without belonging to us? This principle is God, the first and the last principle of everything. Now when the faith of reason in itself is attached to its principle, when it knows that it comes from God, it increases not merely in degree, but in nature, by as much, so to say, as the eternal substance is superior to the finite substance in which it makes its appearance. Thus comes a redoubled faith in the truths revealed by the supreme reason in the shadows of time, and in the limitations of our weakness."†

103. Cousin quotes in illustration, the following passage from Fenelon: *Existence of God*, Part I, ch. 4, *Of Human Reason*. "In truth, my reason is in myself, for it is necessary that I should continually turn inward upon myself in order to find it, but the higher reason, which corrects me when I need it, and which I consult, is not my own, it does not make a part of myself. Thus, that which might seem the most our own, and to be the

* Ps. 119, v. 130.

† Cousin: *El. of Psychology*, pp. 299, 300.

very foundation of our being, I mean our reason, is that which least belongs to us, which we are to believe the most borrowed. We receive continually and at every moment, a reason superior to ourselves, just as we continually breathe an air which is not of ourselves, or as we constantly see the objects around us by the light of the sun, whose rays do not belong to our eyes. There is an internal school, where man receives what he can neither acquire himself, nor learn from other men who live by alms like himself. Where is this perfect reason which is so near me, and yet so distinct and different from me? Is it not God himself, the being for whom I am inquiring?"

104. "There are cognitions in the mind which are not contingent,—which are necessary,—which we cannot but think,—which thought supposes as its fundamental condition. These cognitions, therefore, are not mere generalizations from experience. But if not derived from experience, they must be native to the mind; unless on an alternative that we need not at present contemplate, we suppose with Plato, St. Austin, Cousin, and other philosophers, that Reason, or more properly Intellect, is impersonal, and that we are conscious of these necessary cognitions in the Divine Mind. . . . . On the power possessed by the mind of manifesting those phenomena, we may bestow the name of the Regulative Faculty. This faculty corresponds in some measure to what in the Aristotelic philosophy was called Νοῦς,—νοῦς (*intellectus*, *mens*), when strictly employed, being a term in that philosophy for the place of principles,—the *locus principiorum*. It is analogous likewise to the term *Reason*, as occasionally used by some of the older English philosophers, and to the *Vernunft* (*reason*) in the philosophy of Kant, Jacobi, and others of the recent German metaphysicians, and from them adopted into France and England. It is also nearly convertible with what I conceive to be Reid's, and certainly Stewart's notion of Common Sense."*

105. Probably no English thinker has ever devoted so much attention to the limits and offices of the different Intellectual Powers as Sir William Hamilton, and I note with peculiar satisfaction, the exact accordance of his division of the Cognitive Faculties, both in order of development and relative position, with my own views of the province of Rationality. To show the extent of this accordance, I will quote the closing remarks of his twentieth lecture on Metaphysics.

106. "Such are the six special Faculties of Cognition; 1°, The Acquisitive or Presentative or Receptive Faculty, divided into Perception and Self-Consciousness;† 2°, The Conservative or Retentive Faculty, Memory; 3°, The Reproductive or Revocative Faculty,

* Hamilton: *Metaphysics*, p. 277.

† The only point on which I am inclined to question this division, is the propriety of regarding Self-Consciousness as a faculty collateral with Perception. I would rather view it as a form of Sense or of Intuition, each of which, in its turn, is a form of Perception.

subdivided into Suggestion and Reminiscence; 4°, The Representative Faculty, or Imagination; 5°, The Elaborative Faculty or Comparison, Faculty of Relations; and 6°, The Regulative or Legislative Faculty, Intellect or Intelligence Proper, Common Sense. Besides these faculties there are, I conceive, no others; and in the sequel, I shall endeavor to show you, that while these are attributes of mind not to be confounded,—not to be analyzed into each other,—the other faculties which have been devised by philosophers are either factitious and imaginary, or easily reducible to these."

107. As this division was purely the result of observation and study, the grouping is marked by no law except that of regular gradation, from the form in which the relative antecedence is most objective, to that in which it is most subjective. The Presentative, Conservative, and Reproductive Faculties may all be ranked under Perception,—the Representative and Elaborative Faculties under Judgment,—and the Regulative Faculty corresponds to Understanding.

CHAPTER III.

SUBORDINATE FACULTIES.

108. In seeking suitable terms to designate the secondary faculties (or the mental powers in the third order of our schematic division of Consciousness), we may proceed in either of three ways.

1. By selecting at random names that have been employed by different philosophers, and by a careful analysis of their meaning, assigning their proper place under the motive, spontaneous, or rational forms of Motivity, Spontaneity, or Rationality.

2. By the synthetic addition to each of the faculties that have already been determined, of the peculiar modifications, which may be considered as specially designating their motive, spontaneous, and rational forms, and assigning names that will indicate those modifications.

3. By comparing and combining the respective offices of two or more faculties that have already been ascertained, in order to form an approximate idea of the nature of a faculty that is designated by their united symbols. The symbolic faculty RSM, for example, may be regarded either as (R, SM) the instinctive form of Rationality, or (RS, M) the motive form of Judgment.

109. The farther this process of analysis is carried, the more minute become the distinctions between the several faculties. It is consequently more difficult to find names which

will be precisely and fully significant and exhaustive, and we can hardly hope that our first efforts at nomenclature will be faultless, even if they are anything more than tentative and suggestive,—furnishing a groundwork for the investigations and modifications of subsequent inquirers. But however imperfect our labors may be, the symbols will indicate with great accuracy, the meaning we attach to the terms we employ, and thus furnish to those who follow us, a key to our theories as well as a guide to their own studies, and a means of gradually perfecting the system which we are jointly striving to develop.

110. The result of considerable study and examination, according to each of the above enumerated methods, is the following list of secondary faculties:

MMM, Proclivity,	MMS, Appetence,	MMR, Attachment,
MSM, Selfishness,	MSS, Curiosity,	MSR, Purpose,
MRM, Enjoyment,	MRS, Approval,	MRR, Respect,
SMM, Cautiousness,	SMS, Forecast,	SMR, Constructiveness,
SSM, Attention,	SSS, Direction,	SSR, Resolution,
SRM, Vivacity,	SRS, Concentrativeness,	SRR, Decision,
RMM, Sense,	RMS, Memory,	RMR, Intuition,
RSM, Discernment,	RSS, Deliberation,	RSR, Discursiveness,
RRM, Conception,	RRS, Abstraction,	RRR, Comprehension.

111. The general assignment of these faculties can be made without much difficulty or hesitation, as follows:

Proclivity, Appetence, Attachment, Selfishness, Curiosity, Purpose, Enjoyment, Approval, and Respect are all aroused by directly objective influences, and tend to produce subjective action. They are therefore classed under Motivity.

112. Cautiousness, Forecast, Constructiveness, Attention, Direction, Resolution, Vivacity, Concentrativeness, and Decision, all indicate faculties which are influenced to peculiar modes of subjective action, by subjective motives. They are therefore classed under Spontaneity.

113. Sense, Memory, Intuition, Discernment, Deliberation, Discursiveness, Conception, Abstraction, and Comprehension are all employed with direct reference to objective truth, and are therefore faculties of Rationality,

114. The special place occupied in the general assignment, seems to require notice in a few instances, which will serve to illustrate our several methods of analysis.

115. Appetence may be regarded either as a propensity or a desire, according to the latitude of meaning we accord it. By giving it the symbol MMS, we may indicate this equivocal significance, for it will then represent the spontaneous form of Propensity (MM, S), and the desiring form of Motivity (M, MS).

116. Attachment partakes at once of the nature of Propensity and Sentiment, but its propense is rather more strongly marked than its sentimental character. The symbol

MMR represents the rational form of Propensity (MM, R), and the sentimental form of Motivity (M, MR).

117. Resolution may be ranked without much impropriety, either under Will or Energy. The more common usage, however, seems to imply that a man may have the will to resolve, without the requisite energy to perform, and whatever energy may be embraced in Resolution, seems therefore to be subordinate and not paramount. We therefore regard it as the rational form of Will (SS, R), and the energetic form of Spontaneity (or Spontaneity *tending* to Energy) (S, SR).

118. Concentrativeness might also be classed either under Will or Energy. It resembles the faculty of Resolution, but its character is rather more rational, which is indicated by giving precedence to the symbol of secondary Rationality (S, RS, instead of S, SR).

119. Discursiveness, or the faculty of logical inference, seems to involve the exercise of Understanding, but that exercise is consequent, and not antecedent. Logical reasoning implies comparison and judgment, and its faculty may therefore be properly considered as the rational faculty of Judgment (RS, R).

120. Abstraction, or the faculty which separates the essential part of any idea from what is merely accidental, contradictory, or alien,* seems to involve the exercise of Understanding upon the determinations of Judgment, and its symbol is therefore RRS.

121. Some of the remaining terms are used with such breadth of meaning, that their vulgar acceptation is not precisely defined by the symbolic boundaries which we have given them. This incomplete accordance invites special discussion as to the propriety of our limitation, and the desirability of seeking terms more exact and less ambiguous, for the places they occupy.

122. The office of the several secondary faculties may perhaps be rendered clearer by the following homogeneous classification:

1. Class of pure Motivity.

   MMM. Proclivity. The simplest form of disposition to action, on the presentation of objective impulse.

2. Class of duplicate Motivity and Spontaneity.

* "Abstraction is the selection by the mind of those partial phenomena which admit of being subsumed under one principle. Now it is clearly impossible that this uniting principle should originate in the objects, for another principle would always be necessary in the subject, in order to recognize the unity of the principle in the objects. The principle therefore originates in the subject, and as this is valid of all our observation of phenomena, it follows that the sphere of the application and validity of causality is limited by subjective principles of thought, and cannot be predicated of those things with which the subject has no concern, *i. e.*, of things which are not objects for it at all." *Solly*, p. 69.

MMS. Appetence. Spontaneous Propensity (MM, S), or desiring Motivity (M, MS). A disposition to seek after the simplest form of subjective gratification.

MSM. Selfishness. Motive Desire (MS, M), or instinctive Motivity (M, SM). Spontaneity is more prominent than in Appetence; the disposition to activity is therefore greater, and the subjective gratification that is sought is of a higher order.

SMM. Cautiousness. Motive Instinct (SM, M), or propense Spontaneity (S, MM). The still greater prominence of Spontaneity produces a special reference to the active subject, while the inclination to action (MM) is so feeble, that it may be easily overcome by external obstacles.

3. Class of duplicate Motivity and Rationality.

MMR. Attachment. Rational Propensity (MM, R), or Sentimental Motivity (M, MR). Principally Motive, but implying a certain feeling, or objective tendency.

MRM. Enjoyment. Motive Sentiment (MR, M), or perceptive Motivity (M, RM). The sentiment, feeling, or rational motive is more prominent than in simple Attachment.

RMM. Sense. Motive Perception (RM, M), or Propense Rationality (R, MM). The objective or rational becomes the principal characteristic, to which the motive is subordinated.

4. Class of Motivity, and duplicate Spontaneity.

MSS. Curiosity. Spontaneous Desire (MS, S), or voluntary Motivity (M, SS). Motivity is the most prominent, impelling to the exertion of will for purely subjective gratification.

SMS. Forecast. Spontaneous Instinct (SM, S), or desiring Spontaneity (S, MS). Spontaneity being more prominent, capacitates for action tending to accomplish the ends of desire.

SSM. Attention. Motive Will (SS, M), or instinctive Spontaneity (S, SM). Spontaneity is instinctively attentive, and in the form of attention, the Will tends to specific action.

5. Class of Motivity, Spontaneity, and Rationality.

MSR. Purpose. Rational Desire (MS, R), or energetic Motivity (M, SR). Differing from Curiosity (MS, S) in being less purely subjective, and more objective or rational in its tendency.

MRS. Approval. Spontaneous Sentiment (MR, S), or judicious Motivity (M, RS). Motivity, instead of acting as in Purpose, subjectively for a rational end (SR), acts rationally or objectively for a subjective end (RS).

SMR. Constructiveness. Rational Instinct (SM, R), or sentimental Spontaneity (S,

MR). More spontaneous and less motive than Purpose, with Sentiment less prominent, and Spontaneity more prominent than in Approval.

SRM. Vivacity. Motive Energy (SR, M), or perceptive Spontaneity (S, RM). Differing from Constructiveness as Perception (RM) from Sentiment (MR). Differing from Purpose in the greater prominence of Energy and the subordination of Motivity.

RMS. Memory. Spontaneous Perception (RM, S), or desirous Rationality (R, MS). In Vivacity, Spontaneity has special activity for perceived ends. In Memory, Perception is active for a subjective end.

RSM. Discernment. Motive Judgment (RS, M), or instinctive Rationality (R, SM). More subjective and less prominently motive than Memory; more rational and less instinctive than Constructiveness.

6. Class of Motivity, and duplicate Rationality.

MRR. Respect. Rational Sentiment (MR, R), or understanding Motivity (M, RR). Differing from Approval (MRS), in the substitution of the objective reference for the subjective or spontaneous.

RMR. Intuition. Rational Perception (RM, R), or sentimental Rationality (R, MR). More rational than Respect, and only subordinately sentimental.

RRM. Conception. Motive Understanding (RR, M), or perceptive Rationality (R, RM). Resembling Respect, with Understanding made prominent, and Motivity subordinated. Resembling Intuition, with Rationality made prominent, and Perception subordinated.

7. Class of pure Spontaneity.

SSS. Direction. Will acting of its own accord, to guide towards any desired end.

8. Class of duplicate Spontaneity, and Rationality.

SSR. Resolution. Rational Will (SS, R), or energetic Spontaneity (S, SR). Differing from Constructiveness in the substitution of the spontaneous for the motive, or the energetic for the sentimental element.

SRS. Concentrativeness. Spontaneous Energy (SR, S), or judicious Spontaneity (S, RS). More purely energetic, and more objective or rational than simple Resolution.

RSS. Deliberation. Spontaneous Judgment (RS, S), or voluntary Rationality (R, SS). The objective or rational character becomes the most prominent, and Will operates only for determining the object sought. Judgment is more active and prominent than in Concentrativeness.

9. Class of Spontaneity, and duplicate Rationality.

SRR. Decision. Rational Energy (SR, R), or understanding Spontaneity (S, RR).

Resembling Resolution (SSR), but implying a rational or objective purpose for its secondary action.

RSR. Discursiveness. Rational Judgment (RS, R), or energetic Rationality (R, SR). The logical discursiveness implies the energetic exercise of Rationality, the executive Decision involves the rational exercise of Energy.

RRS. Abstraction. Spontaneous Understanding (RR, S), or judicious Rationality (R, RS). Abstraction differs from Decision, in implying the spontaneous exercise of Understanding, instead of the intelligent exercise of Spontaneity; from Discursiveness in the greater prominence of Rationality, and the subordination of Judgment.

10. Class of pure Rationality.

RRR. Comprehension. The highest exercise of reason, involves the separation from the object under consideration, of all that is accidental, foreign, or non-essential, and the determination of the essential or absolute.

123. In proceeding to assign names to the tertiary faculties (or the mental powers in the fourth order of the subdivisions of Consciousness), the distinctive characteristics become still more minute, and the difficulty of finding precise and definite terms is consequently greater. We will therefore content ourselves with a hypothetical nomenclature, without attempting to define the several faculties more precisely than by the symbolic designation of their meaning. In deciding upon the fitness of each name, the following questions should be asked.

1. Does this name generally express, or may it properly be used to express the precise relations indicated by the symbols that are attached to it?

2. Can those relations be expressed more satisfactorily by any other name?

124.

MMMM, Vitativeness,	M-proclivous;	Propensity-propense;	Proclivity-M.
MMMS, Combativeness,	M-appetent;	" -desirous;	" -S.
MMMR, Amativeness,	M-attached;	" -sentimental;	" -R.
MMSM, Alimentiveness,	M-selfish;	" -instinctive;	Appetence-M.
MMSS, Acquisitiveness,	M-curious;	" -voluntary;	" -S.
MMSR, Ambition,	M-purposing;	" -energetic;	" -R.
MMRM, Self-Esteem,	M-enjoying;	" -perceptive;	Attachment-M.
MMRS, Affection,	M-approving;	" -judicious;	" -S.
MMRR, Adhesiveness,	M-respecting;	" -intelligent;	" -R.
MSMM, Envy,	M-cautious;	Desire-propense;	Selfishness-M.
MSMS, Cupidity,	M-forecasting;	" -desirous;	" -S.
MSMR, Approbativeness,	M-constructive;	" -sentimental;	" -R.
MSSM, Marvellousness,	M-attentive;	" -instinctive;	Curiosity-M.
MSSS, Inquisitiveness,	M-directing;	" -voluntary;	" -S.
MSSR, Eagerness,	M-resolute;	" -energetic;	" -R.

MSRM,	Confidence,.	M-vivacious;	Desire-perceptive;	Purpose-M.
MSRS,	Zeal,	M-concentrative;	" -judicious;	" -S.
MSRR,	Emulation,	M-decisive;	" -intelligent;	" -R.
MRMM,	Content,	M-sensible;	Sentiment-propense;	Enjoyment-M.
MRMS,	Hope,	M-remembering;	" -desirous;	" -S.
MRMR,	Sympathy,	M-intuitive;	" -sentimental;	" -R.
MRSM,	Admiration,	M-discerning;	" -instinctive;	Approval-M.
MRSS,	Esteem,	M-deliberate;	" -voluntary;	" -S.
MRSR,	Taste,	M-discursive;	" -energetic;	" -R.
MRRM,	Generosity,	M-conceptive;	" -perceptive;	Respect-M.
MRRS,	Veneration,	M-abstractive;	" -judicious;	" -S.
MRRR,	Conscientiousness,	M-comprehensive;	" -intelligent;	" -R.
SMMM,	Solicitude,	S-proclivous;	Instinct-propense;	Cautiousness-M.
SMMS,	Vigilance,	S-appetent;	" -desirous;	" -S.
SMMR,	Circumspection,	S-attached;	" -sentimental;	" -R
SMSM,	Frugality,	S-selfish;	" -instinctive;	Forecast-M.
SMSS,	Providence,	S-curious;	" -voluntary;	" -S.
SMSR,	Self-Denial,	S-purposing;	" -energetic;	" -R.
SMRM,	Imitation,	S-enjoying;	" -perceptive;	Constructiveness-M.
SMRS,	Device,	S-approving;	" -judicious;	" -S.
SMRR,	Order,	S-respecting;	" -intelligent;	" -R.
SSMM,	Observation,	S-cautious;	Will-propense;	Attention-M.
SSMS,	Scrutiny,	S-forecasting;	" -desirous;	" -S.
SSMR,	Tact,	S-constructive;	" -sentimental;	" -R.
SSSM,	Activity,	S-attentive;	" -instinctive;	Direction-M.
SSSS,	Management,	S-directing;	" -voluntary;	" -S.
SSSR,	Positiveness,	S-resolute;	" -energetic;	" -R.
SSRM,	Intrepidity,	S-vivacious;	" -perceptive;	Resolution-M.
SSRS,	Pertinacity,	S-concentrative;	" -judicious;	" -S.
SSRR,	Self-Reliance,	S-decisive;	" -intelligent;	" -R.
SRMM,	Frankness,	S-sensible;	Energy-propense;	Vivacity-M.
SRMS,	Alacrity,	S-remembering;	" -desirous;	" -S.
SRMR,	Constancy,	S-intuitive;	" -sentimental;	" -R.
SRSM,	Patience,	S-discerning;	" -instinctive;	Concentrativeness-M.
SRSS,	Perseverance,	S-deliberate;	" -voluntary;	" -S.
SRSR,	Inflexibility,	S-discursive;	" -energetic;	" -R.
SRRM,	Dexterity,	S-conceptive;	" -perceptive;	Decision-M.
SRRS,	Courage,	S-abstractive;	" -judicious;	" -S.
SRRR,	Determination,	S-comprehensive;	" -intelligent;	" -R.
RMMM,	Sensation,	R-proclivous;	Perception-propense;	Sense-M.
RMMS,	Self-Consciousness,	R-appetent;	" -desirous;	" -S.
RMMR,	Apperception,	R-attached;	" -sentimental;	" R.
RMSM,	Suggestion,	R-selfish;	" -instinctive;	Memory-M.

RMSS,	Recollection,	R-curious ;	Perception-voluntary ;	Memory-S.
RMSR,	Retention,	R-purposing ;	" -energetic ;	" -R.
RMRM,	Penetration,	R-enjoying ;	" -perceptive ;	Intuition-M.
RMRS,	Ideality,	R-approving ;	" -judicious ;	" -S.
RMRR,	Affirmation,	R-respecting ;	" -intelligent ;	" -R.
RSMM,	Contemplation,	R-cautious ;	Judgment-propense ;	Discernment-M.
RSMS,	Reflection,	R-forecasting ;	" -desirous ;	" -S.
RSMR,	Imagination,	R-constructive ;	" -sentimental ;	" -R.
RSSM,	Meditation,	R-attentive ;	" -instinctive ;	Deliberation-M.
RSSS,	Comparison,	R-directing ;	" -voluntary ;	" -S.
RSSR,	Calculation,	R-resolute ;	" -energetic ;	" -R.
RSRM,	Discrimination,	R-vivacious ;	" -perceptive ;	Discursiveness-M.
RSRS,	Causality,	R-concentrative ;	" -judicious ;	" -S.
RSRR,	Elucidation,	R-decisive ;	" -intelligent ;	" -R.
RRMM,	Individuality,	R-sensible ;	Understanding-propense ;	Conception-M.
RRMS,	Cognition,	R-remembering ;	" -desirous ;	" -S.
RRMR,	Appreciation,	R-intuitive ;	" -sentimental ;	" -R.
RRSM,	Analysis,	R-discerning ;	" -instinctive ;	Abstraction-M.
RRSS,	Synthesis,	R-deliberate ;	" -voluntary ;	" -S.
RRSR,	Generalization,	R-discursive ;	" -energetic ;	" -R.
RRRM,	Insight,	R-conceptive ;	" -perceptive ;	Comprehension-M.
RRRS,	Sagacity,	R-abstractive ;	" -judicious ;	" -S.
RRRR,	Classification,	R-comprehensive ;	" -intelligent ;	" -R.

125. The subdivision could be carried still further if it were desirablé, but enough has already been done to fully illustrate thé principle of arrangement. If this first essay at arrangement has not been entirely satisfactory in all its minutest details, it may, perhaps, at least compare favorably with any previous one, and it should be remembered that facts of any kind that are "to be examined, ought not to be taken at random, but selected on a principle, and arranged in due order and dependence. But this requires no ordinary ability, and the distribution of things into their proper classes is one of the last and most difficult fruits of philosophy."* A slight, and sometimes hardly appreciable change, in the supposed relative ascendency of the partial characteristics, may remove a faculty from one of the primary subdivisions of Consciousness to another. But whatever doubt may be connected with our imperfect appreciation of the relations, the ideal relations themselves are positive, fixed, necessary, eternal, and the more fully we comprehend the value of all the symbols, both simple and complex, the greater precision will attend all our thoughts and investigations.

126. The accompanying diagram exhibits at a glance, the relations of the several sub-

* Hamilton, *Logic*, p. 399.

divisions of Consciousness. The fourth order of subdivision is omitted, as the nomenclature I have suggested is a wholly experimental one, requiring a long series of careful observations before it will be possible to determine whether it has any value. The faculties of the third order are marked with a note of interrogation, to show that farther study is desirable, to ascertain whether their relative assignment is the best that can be made. It is quite probable that some other order of classification may be more convenient for the lower faculties, but I have thought it would be best to show that the principle of trichotomy may be extended as far as the needs of science may require.

CHAPTER IV.

PHILOSOPHY OF CONSCIOUSNESS.

127. Every man feels that his personality does not consist in any peculiarity of form, feature, or complexion, any more than in the shape or texture of the clothes he wears. He finds his body with its limbs and organs of sense, a very convenient and important instrument for the execution of his plans, and he may take pride in the physical beauty, delicacy, or exquisite finish of that instrument, as he would in the symmetry of a horse, or the superior merit of anything else of which he claimed ownership. But the intelligent self,—the *Me*,—sits apart in such almost inaccessible majesty, that many have been accustomed to look upon it as a kind of mythical somewhat, whose very existence is exceedingly problematical,—a mere resultant, perhaps, of the material and physical organization. Such an opinion is of course based upon the assumption that the material is more patent and intelligible than the immaterial,—an assumption that it may be well to test by a brief inquiry into the character of our knowledge of the nature and qualities of mind.

128. Of the essential nature of mind or matter we know nothing. We can judge of them only by the effects they produce upon us by their properties or attributes. These attributes can be considered as belonging to them only in so far as they are *phenomenal*, that is to say, as they *appear* to our observation. What analogy or connection there may be between the phenomenal attributes and the substantial essence, it is impossible, for reasons that will appear in the course of our inquiry, for us to determine. It is, however, obvious that the phenomena of mind are more closely related to the observing mind, than the phenomena of matter. We ought therefore to know more of mind than of matter, and inasmuch as we know nothing of matter except the effects it produces upon our minds, Bishop Berkely and others have attempted with much force and plausibility of

ELEMENTARY FORMS AS

VARIOUS

COMPREHENSION ? RRR

ABSTRACTION ? RRS

CONCEPTION ? RRM

DISCURSIVENESS ? RSR

DELIBERATION ? RSS

DISCERNMENT ? RSM

UNDERSTANDING RR

RS

JUDGMENT

Rationality

R

CONSCIO

INTUITION ? RM

MEMORY ? RM

PERCEPTION RM

SENSE ? RMM

Spont

ENERGY

SR

DECISION ? SRR

CONCENTRATIVENESS SRS

VIVACITY ? SRM

RESOLUTION ? SSR

SS

INDUCTION ? SSS

D BY CONSCIOUSNESS UNDER ITS
LATIONS.

Thought, . . as Cause and Effect, or as Essential and Accidental. If any such relation between the two be assumed, this is not done in consequence of practical self-observation, and it does not lie in the fact of Consciousness.

132. . . "Should such a relation be assumed upon some other ground than that of self-observation, . . then it appears at first sight, that the two elements, as coexistent and inseparable from each other, must be held to be of equal rank; and thus the inward thought may be as well regarded as the foundation, the essence of the outward perception, which in that case would be the superstructure, the accident, as the reverse; and in this way an insoluble doubt would necessarily arise between the two suppositions, which would forever prevent any final decision respecting the assumed relation. . . But should any one look deeper into the matter, . . inasmuch as the inward consciousness embraces even the outward sense itself; since we are conscious of the seeing, hearing, or feeling, but can by no means, on the other hand, see, hear, or feel our consciousness; and thus, even in the immediate fact, Consciousness assumes the higher place,—such an one, I say, would find it much more natural to make the internal Consciousness the chief thing, and the external Sense the subordinate thing; and to explain the latter by the former; to control and try the latter by the former;—and not the reverse."

133. Either Cousin or Hamilton might have thus discoursed, for each of them maintains the supremacy and efficiency of Consciousness,—the subordination of Sensation,—and the relation of the latter as the chronological, to the former as its logical antecedent. Each rejects the theory of Locke, that all our knowledge is derived from sensation and experience, and acknowledges the transcendency of ideas, which alone render sensation and experience possible.

134. There is much in a superficial acquaintance with metaphysical literature, that tends to discourage the ardent seeker after truth, and to strengthen the vulgar opinion that all philosophical research is foolish and unsatisfactory. The pages of an ordinary Encyclopædia will show that in the earliest historical times, the Brahminical sages taught many of the leading doctrines that characterize some of the most distinguished modern philosophical schools. A cursory perusal of the works of Plato and Aristotle reveals the origin of so much of the variety and profundity of thought that later writers would gladly claim as their own, that one is tempted to exclaim with Solomon, "there is nothing new under the sun," and to believe that in what poor, weak, deluded humanity regards as the most exalted sphere of investigation, it is destined to move in a continual circle, making no real progress, but constantly repeating the ideas and systems of earlier ages.

135. But if our metaphysical reading is more than superficial, much of this discouragement will vanish, giving place to a hope, if not to a full conviction, that the day will come when the science of all science will assume a clearness and a definiteness such as it merits.

Human nature has been the same in all ages, and it is therefore natural that the same questions should continually suggest themselves, and that they should receive the same answers, somewhat modified, perchance, by individual idiosyncrasies. The phenomena of mind are as patent to observation as those of matter, and prior to the days of Bacon, the former were more studied and better understood than the latter. It is true that in neither physical nor metaphysical science had much advance been made for many centuries, but in the entire absence of any system by which the acquirements of one age could be readily communicated to the next, how could any advance be anticipated? A life of investigation, however directed, should doubtless bring to light a mass of valuable truth, which might be recorded for the benefit of future investigators in the same field, but if the record shows no connecting thread of thought, which makes all the details parts of a consistent whole, it will be of little value.

136. The study of isolated facts is dry, dull, tedious, and unprofitable, and even if the facts are arranged according to a merely arbitrary system, though their acquisition may be somewhat facilitated, it will yield but little satisfaction. A lifetime might be devoted to learning a dictionary by rote, but the learner would probably be little more skilful in the expression of his thoughts after his task was ended, than he was at its commencement. And in like manner the student who delves for years in the fertile soil of Greek philosophy, will probably make no further progress than his predecessors, even if he accomplish as much as he would have done had he devoted himself to original and unassisted personal investigation. But let him bring to the study a natural system or plan, based upon eternal and unchangeable ideas,—a plan by which all prior observations may be classified, and every fact may be arranged in its proper place, as an illustration of the Divine Thought,—and it will be strange if he does not find much that had been mysterious, made suddenly intelligible, and difficulties that had seemed insurmountable, suddenly removed. The labors of modern metaphysicians have been principally valuable as tending to develop such a system.

137. [The question has often been discussed, whether the Mind ever loses its consciousness. Though the full consideration of this inquiry would involve an investigation into the substantive nature of mind, and its full solution by us is therefore absolutely impossible, it is worthy of some attention, even if it yields no other result than a determination of the true position and dignity of Consciousness.

138. In a sound, dreamless sleep, we give no external manifestation of intelligence or activity, and on awaking we have no recollection of any train of thought that has occupied our minds. But even in slumber the most profound, a loud noise, any application that causes physical pain, or a sudden change, whether of motion, temperature, position, or other external circumstances, gives manifest evidence that the internal watchman never

slumbers, but is at all times ready to receive any impressions that are sufficient to stimulate the wearied nerves into action.

139. Consciousness has been sometimes regarded as only one of many distinct faculties of Mind, and an attempt has been made to show that mental operations are often carried on without our knowledge.* It has been said, for instance, that when we are absorbed in a train of thought, we may hear a clock strike without being conscious of it,—or we may read long passages aloud without being conscious either of the matter or meaning of the author, or even of the exercise of our vocal organs. There are also authentic cases reported of soldiers marching, and of stenographers reporting evidence and debates while they were asleep, and therefore, it is said, in a state of entire unconsciousness.

140. Such instances, however, are evidences only of forgetfulness more or less entire, and they tend rather to prove that Consciousness is always active even during the greatest physical torpor, than to show that it is ever wholly absent. The mind may be so fully absorbed that we do not hear the striking of a clock, but if we do hear it, we must at the moment know that we hear it, though it may produce so slight an impression that we may forget immediately afterwards whether we heard it or not. So in reading or writing, we must know at the time that we recognize the form of every word, though the words themselves convey to us no idea of their meaning, and leave no trace upon our memory.]

141. As all scientific investigation should exclude from the field of its inquiry everything that is known to be beyond its reach, and everything that is merely speculative, we should, if possible, so limit the terms that we employ, as to confine our researches strictly to the territory that we may reasonably hope to explore. In accordance with this principle, the Natural Philosopher, because he cannot conceive of matter apart from impenetrability, inertia, and extension, defines matter as a substance impenetrable, inert, and extended. In like manner, as we cannot conceive of mind, either as possessed of any of the attributes of matter, or as devoid of consciousness, we may define it as *the form of immaterial substance, which is manifested in Consciousness.*†

142. The acquisition of isolated facts is the earliest, easiest, and simplest form of pro-

* For a good presentation of the argument in favor of this view, see Wayland, pp. 110–118. Rauch says, pp. 110, 115, "Yet Consciousness is not annihilated, but continued as dreams indicate, and as the possibility of awaking at a certain hour sufficiently proves. . . . . . The mind sleeps; it is for a time in a state of unconsciousness, while at the same time it has not in the least lost its consciousness; this has only become latent, or is for a time veiled."

† "We not only feel, but we know that we feel; we not only act, but we know that we act; we not only think, but we know that we think. . . . . Consciousness is this interior light which illuminates everything that takes place in the soul; Consciousness is the accompaniment of all our faculties, and thus to speak, their echo." *Cousin: Hist. of Mod. Phil.*, Vol. I, p. 322.

gress in knowledge. These facts, as soon as they are appropriated, excite curiosity, or a desire to know more. This Curiosity is a stimulus to mental exertion.

143. Influenced by the awakened stimulus, the mind acts somewhat blindly and confusedly at first, yet sufficiently to show that it has in itself an inherent and independent power of action. In the simple exercise of his active powers, without any definite object or aim other than the gratification of a capricious will, the child finds an inexhaustible source of enjoyment.

144. As mental development proceeds, we become conscious of a higher power than that of mere activity,—the power of intelligence, which involves the comprehension of truths, relations, and laws.

145. Hence we are naturally led to the study of Consciousness under three distinct forms of manifestation:

1. As a stimulus to exertion, acted upon by external influences. To this form of Consciousness we have given the name of MOTIVITY.

2. As acting of its own accord, free from any extraneous impulse, and stimulated only by its own conscious Motivity. To this form of Consciousness we have given the name of SPONTANEITY.

3. As operating intelligently for the discovery of truth. This third and highest form of Consciousness, to which Motivity and Spontaneity should be both subservient, we have called RATIONALITY.

146. If this division is admitted as being founded in necessity, or even as being natural or appropriate, it is desirable that the respective limits of the three Conscious Forms should be clearly defined and understood, and that we should carefully avoid attributing to either, an influence which it does not properly possess.

147. We might for instance, naturally suppose that Motivity is subject to the direct supervision of Spontaneity and Rationality, and that therefore it may sometimes be subjective both in origin and tendency. Indeed, it may be asked, if our motives are not thus under our control, how is it possible to believe or imagine that we are in any way accountable for our actions?

148. In answer to this question it may be remarked, that we rarely attach any sense of responsibility to our motives, but only to the acts themselves, so far as they are spontaneous or voluntary. All our motives are good, if they are not allowed an undue authority, and it is a part of the province of Spontaneity, aided by Rationality, to determine the amount of influence that we will accord to each. The question of responsibility does not concern creatures of blind fate or necessity, and it can have no reference to man so long as he is irresistibly impelled by any external force, but so soon as he is able to withstand the impulse, and he begins to deliberate, he becomes accountable. In the majority of

cases we are urged by two or more motives acting simultaneously. We may for instance feel at the same moment, an inclination to physical exertion, a desire to employ that exertion in appropriating something to our own use, and a conscientious conviction that the appropriation would be wrong, because the article desired is already the property of another. No blame can be attached to us for either of these three motives, but if our Spontaneity decides to act in accordance with the inclination and desire, and in opposition to the conscientious conviction, we shall feel self-condemned, and responsible for the improper exercise of Spontaneity.

149. Motivity is blind, involuntary, impulsive,—Spontaneity should be always watchful, cautious, deliberate. Motivity is never subject to our control at the moment of action, or in other words, we can never dictate the character of the motives that shall suggest themselves to us on any given occasion. No man by any effort of will or reason, can make the influence of all his motives such as he would like, or such as according to his rational convictions, it ought to be. If we desire to correct our errors, to reform our habits, or to alter our character, the proper way to effect the change, is by exposing Motivity to such objective influences as will tend to strengthen or weaken its particular manifestations.

150. Spontaneity and Rationality can act on Motivity only indirectly. If Spontaneity is sufficiently strong or deliberate to resist in a single instance, a motive which from indulgence has acquired undue strength, the motive will be weakened, and each new successful resistance will diminish its influence in a greater degree, until finally it will cease to operate improperly. In like manner, a motive that is too feeble, may be made efficient by a strong or deliberate volition, and by habitual exercise, it may be strengthened to any desirable extent. In these cases, Spontaneity does not act on Motivity, but in its proper sphere of restraining our impetuosity, and calling upon Rationality to deliberate between our varying inclinations, it provides a way by which Motivity may receive the proper objective bias.

151. If Spontaneity is weak, so that under the influence of strong Motivity it becomes precipitate, virtuous sentiments and desires will still exist, however feeble, and Spontaneity may strengthen them,—increasing at the same time its own efficiency,—by forming the habit of deliberate circumspection, by avoiding temptation, and by seeking proper employment, suitable associates, and such other circumstantial aids as Rationality may suggest, or the sense of duty may indicate.

152. It is doubtful whether there is ever any action of Spontaneity without a motive. In most cases, indeed, in all cases that involve any question of moral right or wrong, a variety of motives are presented in connection. If they all incline to the same course of procedure, Spontaneity will act instantaneously. If there is a conflict among them,

Spontaneity has the power, and it should exercise the power, of waiting for deliberation. If proper deliberation is used, its final action will always be in accordance with the conscientious motive, and with the convictions of Rationality.

153. In these powers of Spontaneity consists our Freedom of Will. We may not be free to act without motives, or contrary to our motives, or even contrary to the strongest motive. But we have the power of discerning in all cases which *ought to be* the strongest motive,—the power in most cases, of deliberating between conflicting motives, and delaying the too violent (and therefore vicious), until the feebler virtuous motive can assert its supremacy, thus determining which *shall be* the strongest motive,—and the power of exposing ourselves to external influences that will operate on our Motivity for the formation of virtuous habits. For the proper exercise of these powers, we undoubtedly are, as we feel ourselves to be, responsible and accountable.

154. As upon a cursory view, the action of Motivity sometimes seems to have a subjective origin, so may the tendency of Spontaneity often appear objective. Indeed, most of our actions are upon objects,—the very brain that we use as the organ of thought, the nerves that convey our physical volitions, and the muscles that serve as the instruments of our will, being objective when considered in their relations to the mind.

155. But the tendency and end of Spontaneity considered in itself, is merely our personal gratification. Considered simply as active beings, we are conscious of nothing but our activity, which is purely subjective. Rationality alone can take note of any external objects, or declare that our exertions produce any objective effect, and it is only because of the intimate connection of Rationality and Spontaneity,—part of the province of the latter, in aiding its own determinations, being to call for the decisions of the former,—that we could ever suppose Spontaneity to be, like Rationality, subjective-objective.

156. We have said that Spontaneity should be deliberate. By this we mean that Spontaneity should call Rationality to its aid, on every occasion of conflicting motives. Deliberation is not one of the offices of Spontaneity, except inasmuch as it involves the faculty of attention, but under the influence of the conscientious motive that declares the duty of deliberation, Spontaneity may become attentive, and excite Rationality to deliberate. Rationality is the judicial, Spontaneity the executive power. The former expounds and interprets laws that are based upon eternal necessity, and revealed by Divine Benevolence,—the latter, when in the proper performance of its functions, governs all its operations by those laws.

157. Here again we have another apparent tendency inconsistent with our principles. If Rationality decides, and Spontaneity acts according to the decision, it may appear that Rationality tends to determine the exercise of Spontaneity, and is therefore subjective-subjective. But if we reflect, we shall perceive that Spontaneity has determined, before

appealing to the tribunal of Reason, to act in accordance with the motive which ought to be the strongest, according to the convictions of Rationality. Rationality makes its objective decision on the questions propounded to it by Spontaneity, and its office is then accomplished. Spontaneity makes use of the decision, and if its determination has not been changed in the meanwhile by a new intervention of Motivity, calling for a new exercise of its powers, it makes that the strongest, which might otherwise have been the weakest motive.

158. The immediate origin of Rational action sometimes appears objective. In perception through the senses, if we regard Perception as a rational faculty, it is not always easy to perceive any intermediate action of Motivity or Spontaneity, between the physical impression upon the nerves, and the intelligent perception. But if we examine closely, we shall probably find that in every instance, the spontaneous faculty of attention is aroused, before any perception can take place.* If the attention is wholly absorbed, pictures of passing objects may be painted upon the retina in the most glowing colors, the waves of sound from the most soul-thrilling melodies, may beat their tattoo on the drum of the ear, the pores of the sensitive skin may be closed by cold, or opened by sweltering heat, without exciting perception, the impression upon the consciousness being insufficient to divert the action of Spontaneity, so that Rationality may assign to the impression an objective validity. If amidst this absorption, we are suddenly startled,—as for instance, by a vivid flash of lightning, the near report of a cannon, or a violent blow,—we have first a confused consciousness of disturbance, to which succeed a motive desire to understand the cause of the disturbance, and a spontaneous act changing the direction of the faculty of attention, followed immediately by a rational objective perception.

159. In every instance of the recognition of a physical object, the process appears therefore to be,—first, an impression on the brain through the nerves,—second, if this impression is sufficiently strong, a simple and at first confused consciousness of that impression, exciting Spontaneity through the intervention of Motivity,—third, a rational perception of an object. Between the objective impression and the objective determination of Reason, an objective-subjective act must invariably and necessarily be interposed, otherwise the action would be merely objective-objective, and as such it would be entirely excluded from the sphere of Consciousness.

160. Because perception follows the impression on the sensitive nerves, without any apparent interval, it is not strange that we should think it impossible for any other mental

* St. Jerome, quoted by Sir William Hamilton, Reid's Works, p. 877, says: "Quod mens videat et mens audiat, et quod nec audire quidpiam nec videre possumus, nisi sensus in ea quæ cernimus *intentus*, vetus sententia." Sir William Hamilton makes "an act of *Attention*, however remiss," the first condition of perception.

operation to intervene. But when we observe the rapid movement of the fingers in a skilful pianist, or the marvellous facility of computation in an experienced accountant, and reflect that each motion, and each addition, requires a distinct volition, we may readily conceive that Motivity and Spontaneity may have time to act between the consciousness of the sensible impression, and the perception of the object from which that impression originated. And if we watch the earliest developments of the perceptive powers in an infant, we can hardly fail to be convinced that a desire to interpret the unknown affection of Consciousness, precedes every distinct perception. If we find such a precedence in but a single instance, it will furnish a strong presumption that the order we have indicated is the natural one, and that it is only because it is habitual, that we fail to detect it in every case.

161. However philosophical the division of Consciousness into Motivity, Spontaneity, and Rationality may be, as a basis for the classification of the mental faculties, the division is one, as we have seen, that can never fall purely under our observation, but it is rather a rational determination *a priori* of necessary states, like our conception of matter, which is derived from the consideration of mixed and varying phenomena or qualities. Our ideas of the three Conscious-forms may however be made more distinct, definite, and adequate, than our ideas of matter or its primary attributes, and therefore the science of mind has a more impregnable foundation than that of matter. Whether upon that foundation a metaphysical superstructure will ever be erected, more beautiful and complete than our present congeries of physical sciences, is a question for the future to solve.

162. Consciousness is so far a rule to itself,—the mind is so multiform in its infinite variety of capability, that we might plausibly demand even more latitude in our attempts at defining and comprehending its divers characteristics, than we so readily allow to every student of the more precisely marked forms of less versatile material nature. We have however no occasion for any such demand, for the science of mind, at least in its foundation, is not only more precise and definite, but it is also more substantial than that of matter. We use the term substantial in its primitive meaning, as indicating an approximation to the perception of that which underlies the phenomenal or accidental. From the necessary relations of the subjective and objective, we have already deduced the three classes of mental activity, which correspond to Motivity, Spontaneity, and Rationality. Does this deduction fail in any respect, of being exactly and rigorously scientific? The distinction into the subjective and objective is real and definite,—the four classes which indicate the progress of activity from its origin to its termination, are exhaustive and positive,—the first of these four classes (the objective-objective), is of course excluded from any connection with the subjective, the three classes remaining are distinctly characterized, and harmonize wonderfully in their significance with the three modes of mental development. What more need we desire?

CHAPTER V.

KNOWLEDGE AND FAITH.

"Are those who would make man the measure of all things, sure that they have found man's true measure?"

The Patience of Hope: Boston ed., p. 84.

163. THE object of every science is the discovery of truth.

164. In many investigations, as for instance in the propositions of Geometry, we arrive at results which it is impossible to doubt,—results which are recognized as necessary by every one who can understand the train of reasoning by which they were obtained. The argument which conveys this necessary and universal conviction, is called demonstration, and the sciences which are built upon demonstration are called exact sciences.

165. But in the inquiries which are of the most general interest, such as the character of our own being, our relation to our fellow-men and to the universe in which we are placed, our duties and the proper mode of determining them, conflicting opinions are held and zealously maintained. Demonstration of mental and of moral truth seems unattainable, and if we seek for illumination from the writings of metaphysical philosophers, we can hardly fail of being led into skepticism or universal doubt, and we may esteem ourselves fortunate, if we are able to pass over the abyss of doubt into settled conviction, or even into partial belief.

166. Two and two make four. The three angles of a triangle are equal to two right angles.

167. These two propositions are mathematical truths, the first being an axiom or self-evident proposition, the other a theorem or demonstrable proposition. Every intelligent person accepts them without hesitation, and without requiring a strict definition of all the terms employed. The meaning of two, three, and four, of angle, right angle, and triangle, is supposed to be sufficiently obvious, and such questions as "How do you know that two is always two?" "How do you know what an angle is?" would be generally regarded as indications of impertinence, or insanity.

168. Every effect must have a cause. God is. The soul is immortal.

169. These three propositions are all metaphysical. We may not perhaps be able to decide upon their truth or falsity, until we have obtained satisfactory answers to questions like the following. What is an effect? What is a cause? What is God? What is the soul? What is immortality? How do we know that an effect must have a cause,—that there is a God,—that the soul is immortal?

170. There is, then, a difference in the character of propositions that may be presented for our consideration. In what does this difference consist? Why do we require a greater amount of information, and a closer investigation in one instance than in another? What are the characteristics of truth, and on what authority can we rely as the arbiter of certainty? What is the nature of fundamental truth, and in what manner should we proceed to increase our knowledge, by the comparison of fundamental truths?

171. The most obvious knowledge is that which is purely sensual.

172. Whatever views we may hold with regard to our spiritual nature, or the object of our being, we cannot overlook our intimate connection with the material universe. That connection is maintained and recognized by the five senses, which we possess in common with many of the inferior animals. The evidence of the senses is purely personal. We neither require any higher authority than ourselves to decide whether we really see and feel, nor can we admit any right or possibility inherent in any superior being, to give validity to our sensations. If a ball is placed in my hand, I may question as to its nature, the nature and attributes of matter, the relation which exists between the ball and my perceptions, the mode in which a knowledge of its existence is conveyed to my brain, but I cannot doubt,—in other words, I know that I see and feel,—that there is a something, —call it matter, force, spirit, or whatever you will,—that produces sensations, the aggregate of which I define by the term sphere.

173. Among the most obvious ideas suggested to us by the senses of touch and sight, are those of form and proportion. The idea of proportion is also conveyed by the ear, since all harmony requires that a determinable mathematical relation shall exist between different vibrations. We have also other sensual ideas, such as those of light, color, heat, taste, smell, between which it is more difficult to discern any general connection. But even in some of these the laws of proportion are traceable, and if the undulatory theories of light and heat are correct, nearly all the impressions of our senses may be subjected to mathematical calculation.

174. Mathematics may be defined as "the science of form and proportion," proportion including the idea of number. Its demonstrations are obtained by observing the relations between certain axioms, or self-evident truths. The relations, as well as the axioms themselves, must be self-evident, and they are self-evident because all sensual perceptions are self-evident, there being no tribunal conceivable higher than ourselves to which they can be referred for decision.

175. It is true that most of the propositions of pure mathematics are abstract and general. There is therefore a mental effort superadded to the sensible impression. But this effort is made by ourselves, in our own right, and constitutes a part of the judgment from which no appeal is either possible or desirable. Thus the abstract idea of two, is

part of the impression conveyed to the mind by the sight of two objects, and we can employ that idea in any train of reasoning, with the same unhesitating confidence as we give to the simple original perception of two distinct objects.

176. From these considerations we may discover a sufficient reason for the implicit faith that we place in mathematical axioms. Without stopping for the present to inquire whence the power is derived, we know that there is a power inherent in our own nature, by which we perceive their truth. They constitute a part of our immediate perceptions, and each individual is necessarily the only judge of what he himself perceives.

177. It is often said that our senses deceive us. Is this assertion true, or is it our judgment that deceives us, and are we led into error by a hasty or improper exercise of our own powers?

178. Let us suppose the following question propounded to a person of sound faculties and mature judgment, but one who is entirely ignorant of chemistry, and of the results produced by the mixture of different ingredients.

"If I were to mix two quarts of one fluid, with two quarts of another fluid, how much would there be in the whole?"

The answer would probably be, "Four quarts."

"How do you know that there would be four quarts?"

"Because two and two always make four."

179. But it could be easily shown, by mixing two quarts of sulphuric acid and two quarts of water, that in consequence of the chemical affinity existing between the liquids, a condensation would take place, so that there would be less than four quarts of the mixture. Whence then did the error of opinion arise?

180. Certainly not from the mathematical axiom, for our confidence in its truth would still be unshaken, but from a hasty judgment, and from losing sight of the precise meaning and extent of the axiom.

181. Suppose again the following conversation with a man well versed in plane geometry, but entirely ignorant of the nature of spherical triangles.

"To what is the sum of the three angles of a triangle always equal?"

"To two right angles."

"Would it be possible to construct a triangle, in which the sum of the angles would be either greater or less than two right angles?"

"It would not?"

"Are you sure of this?"

"I am, as sure as I am that two and two make four."

"And yet, as I will show you upon this sphere, we may describe a triangle, in which the sum of the angles shall be nothing, another in which it shall be equal to six right angles,* and others in which it shall be equal to any quantity we please, from 0° to 540°."

182. This conversation and experiment would not weaken the belief in the truth of the original proposition, as it had always been understood. It would merely show that the judgment had assumed too much, or that the definition of the term "triangle" was too limited, and that the proposition was true only of rectilinear plane triangles. It would also show that demonstrable truth may lead us into error, if it is not perfectly understood, and if its full extent and limits are not properly recognized. Hence the seeming paradox, that a thing may be proved true, though it is absolutely false.†

183. Of a similar character are the errors which we attribute to the senses. The nerves connected with each organ of sense are designed to convey certain appropriate

* This statement is true only in the sense in which the Calculus disregards differentials. In order that there may be an angle at the junction of any two sides of a spherical triangle, each of the three angles must be less than 180°, but it may differ from 180° by a quantity less than any assignable value, therefore it may virtually be regarded as equal to 180°. The algebraical fallacy in the following note, shows that it is not always safe to disregard differentials.

† The following algebraical demonstration that 1 is equal to 3 will illustrate my meaning.

$$\text{Let } x = a$$

$$\text{Then } x^2 - 2ax + a^2 = x^2 - 2a^2 + a^2 = x^2 - a^2$$

$$(x - a)(x - a) = (x - a)(x + a)$$

$$x - a = x + a$$

$$x = x + 2a = 3x$$

$$1 = 3$$

The error in this case consists in assuming that any factor which has no value, can be used as a factor in determining the numerical value of a quotient.

Peter Barlow (*Elementary Investigation of the Theory of Numbers*, London, 1811, Prop. IX), demonstrates that "the sum of any number of prime numbers in arithmetical proportion, is a composite number." He overlooked the arithmetical series 1, 2, 3, in which 1 + 2 = 3, and 2 + 3 = 5,—both 3 and 5 being prime numbers. With this exception, the demonstration is perfectly rigorous.

Prof. Pierce (*Mathematical Monthly*, October, 1858), gives a number of "Propositions on the Distribution of Points on a Line," all of which are rigorously true in their intended meaning; but in some of the cases, it is necessary either to suppose that the line is straight, or that the distances between the assumed points are measured *on the line*, and not in the direction of one point from the other.

Such instances in the "exact" sciences, teach the necessity of a precise understanding and exposition of the principles, as well as of all the relations involved in any train of reasoning. The very possibility of error is a proof both of liberty and of imperfection.

We can reason only about that which we can define, and we can define any proposition only as it is comprehensible to us. All contradictions and errors can probably be traced to errors of definition. It may often be seen by impartial observers, that two disputants are both right, and that they differ only because each does not see the phase of truth at which the other looks.

sensations to the brain. If I press upon the ball of my eye, the optic nerve will convey the impression of light to which it is adapted. If the bloodvessels of the head are unnaturally distended, so as to compress the auditory nerve, that nerve will also convey the only impression of which it is susceptible, that of sound. There can be no possible doubt of the reality of the impressions or sensations, but there may be a doubt as to the cause of those sensations. If the judgment is made hastily, and without due regard to all the circumstances which ought to be taken into consideration, it will probably be erroneous.

184. The healthy optic nerve not only perceives the rays of light, but also the direction in which they come. The mind perceives that distant objects are more indistinct than those that are nearer. Therefore, if anything is seen which subtends a small angle, but is very indistinct, we may naturally suppose that it is a large object; whereas, if it had been perfectly distinct and subtended the same angle, we should have judged it to be very small. Our judgments formed in this manner may be generally correct, but they will not be infallible, unless the cause of the indistinctness is perfectly understood.

185. If we could conceive that any object at which we were looking actually touched the eye, we should think it exceedingly diminutive. If we supposed it to be within the eye, at the intersecting point of the rays from the top and base, it would seem to be a mere point; and if we could possibly fancy that it was near the retina, we should believe that it was reversed.* There are, therefore, various ways in which a judgment, based either on the evidence of our senses, or on mathematical axioms, or on propositions demonstrably true, may be deceptive, but in every instance our error will be found to arise from a partial or improper use of some of our faculties.†

186. All deception is a virtual lie.

187. If I place in the hands of a pupil a book in which the words are so far perverted from their usual meaning that he cannot fail of receiving a false impression, and if, al-

* Prof. Liedenfrost's case of the young man who first received his sight when he reached his seventeenth year, and to whom all objects at first seemed inverted, can be easily explained on this principle. See *Wayland*, p. 76.

† A great deal of needless obscurity has been thrown by some writers about the subjects of erect and binocular vision. If we looked merely at the images on the retina, we could not fail of seeing two images, both inverted. But the simple hypothesis that we look at the objects themselves, and that the eye informs us correctly of the direction of the luminous rays that proceed from all parts of the object, is not only entirely consistent, but it removes all difficulty. Then, if our judgment fixes the relative position of the object accurately, its size, outline, and solidity will be determined with mathematical precision.

It would be well for philosophy to get rid of the idea of images, as entities distinct from the objects themselves. Whether the rays of light come to the eye from a reflecting, or through a refracting medium, they come from the object that is seen; and it is as proper for us to say that we see *ourselves* in a mirror, as to say that we see a star through a telescope, a stone under water, or a cloud in the air. In each case the rays of light are diverted from a direct course, and it is the office of judgment to determine the extent and cause of the diversion.

though conscious of that perversion at the time, I do nothing to correct the false impression, I deceive him,—I lie to him; and my guilt is as great as though I had communicated the falsehood to him verbally.

188. But if I merely give him a work written in a foreign language with which he is somewhat familiar, and if he falls into error through carelessness in consulting his lexicon, or in the use of a faulty grammatical construction, the error is no longer mine. I have been truthful, but he has been guilty of a mistake. The mistake has arisen from an improper use of his free agency.

189. We are all pupils in the school of the universe.

190. The Power that gave us being has created us with a certain physical organization and certain spiritual faculties, by means of which we are connected with the physical and spiritual world. If the natural and proper exercise of any one of the faculties leads us into error, the responsibility of that error rests with the Creator, but if we are deceived by an improper use of our powers, we alone are responsible.

191. All that is self-evident is, therefore, true. All truth is a revelation from God. Revelation is perfect and continual. It is not confined to mere words, times, or localities. It is uttered in a language that all can understand, at all times and in all places, where a Soul is found capable of receiving it. It comes in music to the ear, in beauty to the eye, in symmetry to the touch, in perfume to the smell, in pleasant savor to the taste, in truth to the mind. It is independent of human agency and human laws, its truthfulness depending on the highest conceivable authority, the word of the Almighty.*

* Some objection may perhaps be made to the use of the term Revelation in this broad sense, but I know of no other term that will so well express the "unveiling" of eternal and necessary truths. Since the days of George Fox, the belief has become general among different denominations of Christians, that our conscientious convictions of duty are immediately revealed to us by the Holy Spirit; and as it cannot be doubted that our fundamental beliefs are implanted in us by the Creator, I can see no impropriety in classing them with other and higher revelations from the same authority. He who most fully recognizes the indubitable character and Divine origin of the faith on which all his knowledge rests, will be best prepared to perceive that reason without faith is a delusive guide, and that the revealed records contained in the Holy Scriptures, appeal to a higher and more authoritative portion of our spiritual being than the logical faculties,—in other words, that Faith is higher than Reason.

St. Augustin and Luther speak of our primitive beliefs as acts of faith,—Reid, Stewart, Degerando, Jacobi, Cousin, and others, call them revelations or inspirations. See *Reid*, pp. 760–1.

"That philosophy is the only true, because in it alone *can* truth be realized, which does not revolt against the *authority* of our natural *beliefs*.

'The voice of Nature is the voice of God.'"

"Consciousness is to the philosopher, what the Bible is to the theologian. Both are professedly revelations of Divine truth." Hamilton, *Discussions*, pp. 69, 90.

"Let every good and true Christian understand that truth, wherever he finds it, belongs to *his* Lord. . . By

192. Imperfections of language, dulness of comprehension, hastiness of judgment, argue no defect in our powers of perception. In explaining to others a truth that we clearly perceive and know, we may inadvertently convey a wrong impression, in consequence of a want of precision in our words, but we do not thereby detract from the truth *as it is perceived in our own minds.* To take one of the examples already adduced: if I state the proposition that the three angles of a triangle are always equal to two right angles, my neighbor, from his confidence in my knowledge of the science of geometry, may believe my statement, and may erroneously suppose that the proposition is true of spherical triangles. The idea, however, as it exists in my own mind, being founded on self-evident relations between self-evident propositions, is incontrovertibly true, the error arising from the fact that my neighbor embraces spherical triangles in his definition of triangles, while I do not.

193. His error is one of mere belief, not of knowledge. He would hardly say that he *knows* the sum of the angles of a spherical triangle is equal to two right angles, and if he should say so, he could assign no better reason than his confidence in my assertion, an assertion that he misunderstood, and he would thus show that by the term knowledge, he merely meant confident belief.

194. All error is merely of belief. It is always based on truth, and in a certain sense, may be said to represent partial or possible truth. It would be adopted by every mind that reasoned from the same data, for it may be laid down as a law of our nature, that if a series of facts or arguments be presented in precisely the same order, under the same circumstances, and with the same degree of relative strength, to two different individuals, they will both deduce the same conclusions. But if one perceives any relations which are obscure to the view of the other, a difference of opinion will immediately arise.

195. Let us briefly recapitulate the postulates we have endeavored to establish.

196. All our faculties are implanted in us by the Creator. Every opinion that is formed necessarily and irresistibly, from the use of those faculties, must be true, and may be regarded as a revelation from God, with as much propriety as if He verbally assured us that it was true. Therefore every proposition that is self-evident, or that is traceable through a series of self-evident relations to one or more axioms is true, and constitutes a

whomsoever truth is said, it is said through His teaching, who is the truth." St. Augustin, quoted by Butler, Vol. II, pp. 43–4.

"The objections made to Faith are by no means an effect of knowledge, but proceed rather from ignorance of what knowledge is." Bishop Berkeley, quoted by Mansel.

"It is not improbable that the writings of Proclus were indebted to Christianity for a term that occurs with peculiar frequency in them,—the term πίστις, or faith, which Proclus regards as direct communion with the Infinite and Absolute, and the highest faculty of the human soul." Butler, Vol. II, p. 330.

portion of our absolute knowledge. But every proposition that is based either wholly or in part, on data that are not fully comprehended, or on ideas the necessary truth of which is not clearly evident to our own minds, may or may not be true; it can only form a portion of our belief, and our belief will be stronger or weaker, in proportion to the number of self-evident truths, or of mere probabilities, that enters into our chain of reasoning.

197. The revelation or immediate judgment is always necessary and infallible, but partial,—expressing a decision only on the premises that are laid before it. The ultimate judgment is more subject to our control by study and care. The man who has fully investigated all his premises, as well as all their relative bearings, will be less in *error* (so far as his positive knowledge is concerned), even though his conclusions may be *farther from the truth*, than those of another who decides hastily and without investigation, but rather from prejudice. This thorough investigation of all the details of our belief is impossible. We are necessarily and properly compelled to place great reliance on authority, and "we should have so much faith in authority, as shall induce us to repeatedly observe and attend to that which is said to be right, even though at present we may not feel it to be so."*

198. The perception of spiritual existence has generally been regarded as one of the characteristics that distinguish man from the lower animals.†

199. If this perception is in reality a distinctive mark of human nature, it should be possessed most highly by those who have the highest spiritual culture.

200. In each faculty of our merely animal nature, there are many of the lower animals which surpass us. Acuteness of vision, quickness of scent, readiness of hearing, are qualities that mark birds and beasts of prey, rather than man. But superior skill, judgment, the capability of indefinite mental development, intellectual and reasoning power, belong to man, if not exclusively, at least in a higher degree than to any other animal.

201. Every faculty, sensual or spiritual, is susceptible of culture. The trained hound will follow the scent of game more steadily than one that is wholly unused to the chase; the educated musician is more sensitively alive to the slightest discord, than the tiro; the thorough mathematician will immediately detect an error of demonstration that would escape the notice of an elementary student.

202. Which of our faculties are the most fully developed, the most diligently and thoroughly educated?

203. The senses and the perceptive faculties whose principal office is merely to take

* Ruskin; *Modern Painters.*

† Solly says (p. 8), "It is the essence of an intellectual nature, to be able to convey its results only to a similar intellectual nature." This is one of the many postulates that lead to the recognition of a Supreme Intelligence.

cognizance of the sensual ideas, are earliest called into action, and during the whole life they are incessantly employed. We see and feel and hear at all times, and nearly at all times our thoughts are engaged with what we see or feel or hear. With the mass of mankind, how small a portion of mental activity is devoted to the consideration of subjects not immediately connected with the daily routine of business. Even the professed student cannot divest himself of any portion of his corporeal nature;—he cannot often even feel that the spiritual maintains an ascendency over the physical.

204. The animal man therefore becomes fully developed by constant exercise; the spiritual man is developed only by a casual and interrupted education. Sensual or animal ideas are therefore the most familiar, and the most perfectly understood; spiritual ideas are but imperfectly comprehended, and are generally wrapped in obscurity and doubt.

205. This is an evident cause of difference in the character of simple propositions. Whatever is discerned by the senses, or by the perceptive faculties through the medium of the senses, is so familiar as to be self-evident, and as we readily see that we are the only possible judges of what we ourselves perceive, we never think of questioning physical axioms. But whatever is spiritually discerned, being somewhat obscure and strange, we are led to question not only the entire perception, but to scan every point that has any connection with it, and to seek for some authority out of ourselves, for that which hardly seems to be a part of ourselves. That authority, as we have already seen, is our Creator, but we can appeal to His authority only for that which becomes to us individually self-evident. Spiritual knowledge, therefore, can only be possessed by those whose spiritual culture approximates nearly to the ordinary physical culture of mankind, but spiritual belief or faith may be cultivated by every one.*

206. It is evident to every student of mathematics, that the theorems and axioms bear such a relation to each other, that if the theorems be assumed true the axioms can be demonstrated. We might readily conceive that beings with a higher order of intelligence than our own, would view the most abstruse propositions of geometry as simple axioms, and we might possibly imagine a mind so constituted that our ordinary theorems would be self-evident, while our ordinary axioms would require demonstration.

207. These supposable cases we find actualized in the study of mind. So different is the constitution of different individuals, that a spiritual truth may be self-evident to one

* "The moral sense indicates that which is above itself, and beyond itself; therefore, if it be our rule to follow always the course of thought, we must now go forward at this suggestion, and it leads us directly to the conception, however vague, of AN AUTHORITY to which we are related. This conception, under all imaginable distortions, has accompanied human nature,—invariably it is the instinctive belief of man. . . . The idea of AUTHORITY, or of a relationship between two beings, each endowed with intelligence and moral feeling, supposes that the *will* of the one who is the more powerful of the two has been in some way declared." *Taylor: World of Mind*, p. 94.

that seems doubtful to another, while other truths may be clearly perceived by the second that are very obscure to the first. For this difference of perception there seems to be no remedy, and hence it is impossible to frame any system of mental science that shall commend itself equally to all.

208. But if each individual would study carefully the workings of his own mind, he would probably find that the same extent of truth is attainable in metaphysics as in mathematics, and that all the great problems of our spiritual nature, and of spiritual existence in general, are as indubitable as the simplest propositions of geometry, provided we pursue our investigations with the requisite diligence.*

209. To one man it may be self-evident that there can be no effect without a cause, and it may also be self-evident that intelligence is the highest possible cause. If he compares these two axioms, he will demonstrate to himself, from their self-evident relation, the existence of an intelligent God, and the demonstration will be entirely and mathematically rigorous.

210. To another man the existence of an intelligent Deity may be demonstrable by the comparison of other truths, which are his axioms, while to a third, the Divine existence may itself be axiomatic. But it can hardly be expected that the train of reasoning adopted by either of the three, will be satisfactory or conclusive to the others, for that is self-evident to one, and therefore incapable of proof, for which the mental constitution of another requires demonstration.

211. The investigation of spiritual truth is therefore full of intricacy. The best spiritual guides are but imperfectly acquainted with the way in which they would lead us, and we must consequently learn to place dependence on ourselves, or rather on the revelations that may be made to us. Self-evidence, as we have seen, is the only test of knowledge, and whenever any truth, simple or complex, becomes self-evident, or is the self-evident result of various self-evident relations, we shall feel that no one can deny it. If any profess to disbelieve it, we shall know that they do not comprehend it, and if any chance to assert that they know it is false, we shall know that they do not understand our meaning.

* Cousin gives the following as instances of metaphysical axioms. "Quality supposes a subject, succession supposes time, body supposes space, the finite supposes the infinite, variety supposes unity, phenomenon supposes substance and being." *Hist. of Mod. Phil.*, Vol. II, p. 283.

CHAPTER VI.

CHARACTER AND LIMITS OF BELIEF AND CERTAINTY.

212. We must be careful to distinguish between *knowledge* and *belief*, for whatever we *know* can never be denied by any one else, but our *belief* may be modified by errors that have been inadvertently admitted. Knowledge is uniform,—belief is manifestly various; the knowledge of one age can neither be falsified nor weakened by the discoveries of a subsequent age, but systems of belief have their rise and fall, and are constantly undergoing modification.

213. Is such knowledge as I include in my definition possible? Are there any facts which we can assert with absolute certainty,—without reference to our own constitution, or the constitution of things around us,—truths independent of all accidental circumstances, independent even of the power of Omnipotence, necessary, indubitable, incontrovertible? Most assuredly there are. We know that we have an existence, a personal being,—that we have certain sensations, thoughts, impulses,—that there is an existence exterior to ourselves, exerting an influence upon us, and capable of being influenced to some extent by us. If we are asked *how* we know all this, we can only answer that we know it,—that we have a faculty given to us by our Creator, which can perceive truth, and know it to be truth. Could that faculty receive any greater authority than it already has? We feel that it could not. We know that it could not. We may prove that it could not. Even without paying any regard to the source of its authority, we see that if any higher tribunal should attempt to strengthen our conviction, it must do so by appealing to this very faculty. In order to have any confidence in the teacher, we must *know* that he is authorized to teach; the final appeal is therefore necessarily to ourselves,—to our own power of knowing.

214. We are apt to confound certainty with demonstrability,—to think that some doubt attaches to all that cannot be proved. We hear much said, and deservedly said, in favor of mathematical science, and the rigorous exactness of mathematical reasoning.* We

* On the other hand, many writers have disparaged the study of mathematics. Hamilton says (*Discussions*, pp. 267–312), "If we consult reason, experience, and the common testimony of ancient and modern times, none of our intellectual studies tend to cultivate *a smaller number of the faculties, in a more partial or feeble manner than mathematics*. . . . The first authority is that of *Bernhardi*, one of the most intelligent and experienced authorities on education to be found in Prussia.

"'It is asked, *Do mathematics awaken the judgment, the reasoning faculty, and the understanding in general*

even meet with those, who claim for mathematics the honor of being the only branch of human knowledge that deserves the name of a science, and we may learn in history, that eminent philosophers have attempted to establish their systems on a mathematical basis,—systems of faith,—systems of religion,—systems of ethics,—systems of jurisprudence. We may even ourselves, have sometimes regretted that subjects which seemed to us of the most vital importance, should be veiled in obscurity, and we must therefore rest our confidence in them, solely on the authority of others. We may have longed for more light,—for a more confident belief,—for certainty, or at least an approximation to certainty.

215. But notwithstanding this longing,—this earnest cry of all humanity for more positive knowledge,—the world still believes that the relations of form and proportion, are the only ones that are susceptible of a rigorous and satisfactory analysis. Whatever its origin, this idea is certainly a mistaken one.

216. Even in mathematics, all things are not proved. The very idea of proof is merely the idea of some new truth, deduced as a necessary inference from the relations existing between other truths which had been previously recognized. If we were obliged to prove everything, we could prove nothing, for we could never have a starting-point. We must therefore recognize certain truths as self-evident; we have already discovered that there are such truths, truths that form the substratum of all our knowledge, and of all our belief,—the axioms of science. These axioms are not confined to mathematics, neither have mathematical axioms any greater certainty than any other self-evident truths. The axioms of our own existence, and of the existence of something independent of ourselves, have no mathematical characteristics, and yet they are as indubitable as the axiom that the whole is greater than a part. Every necessary conviction of the mind, every proposition that we receive unhesitatingly as true the moment we comprehend it, in other words, every self-evident truth, is incontrovertible, and every science that can be built upon such truths, and upon a correct perception of the relations that subsist between them, is a true science, and constitutes a portion of the absolute knowledge to which we are all capable of attaining.

217. There is, then, in the nature of things, no reason why we may not have metaphysical sciences, as well as physical sciences,—sciences of mind, of morals, and of law, as

to an all-sided activity? We are compelled to answer, *No;* for they do this only in relation to a knowledge of *quantity*, neglecting altogether that of *quality*. Further, *is this mathematical evidence, is this coincidence of theory and practice actually found to hold in the other branches of our knowledge?* The slightest survey of the sciences proves *the very reverse;* and teaches us that mathematics tend necessarily to induce that numb rigidity into our intellectual life, which pressing obstinately straight onward to the end in view, takes no heed or account of the means by which, in different subjects, it must be differently attained.'" Von Weiller, Klumpp, Goethe, D'Alembert, Descartes, Du Hamel, Arnauld, and others, are quoted in further illustration of our author's views.

well as sciences of number, of matter, and of form. We may, with as much reason, hope to discover valuable truths by observation and experiment in one case as in another; in the study of any subject that has never been investigated, as in following the beaten track of investigation. But where shall we seek for these truths, and how shall we build upon them after we have laid our foundation?

218. Our earliest knowledge, and, in the opinion of a certain school of philosophers, our entire knowledge is obtained, either directly or indirectly, through the medium of the senses. Long before we are able to express our thoughts, even before we know what it is to think, these busy observers are at work examining the objects around us, and storing our minds with the results of their examination. We are required by our very nature to place confidence in the information that they convey, and that confidence is never materially weakened by the experience of life, or by the arguments of theorists, who tell us we should not depend on the testimony of the senses, because they so often deceive us. Before we reflect at all on the distinctions of truth and falsehood, the sensual impressions have become indelible; they constitute, in fact, a part of our very being,—a reality that we can no more deny than we can deny our own existence. It is evident, therefore, that we can have no assurance with regard to any portion of our experience or belief without relying implicitly on these early impressions.

219. That implicit reliance we all have; it is necessary, irresistible. No arguments ever have been adduced, and we may be assured that none ever can be adduced, to weaken it. We know that the testimony of our senses is true. We know that there is something without us that is capable of exciting certain impressions within us,—that there are real existences, certain qualities of which are cognizable by the sight, the feeling, the hearing, the smell, the taste. What is the nature of external objects, we have no means of knowing, and with our present faculties we shall probably never be able to ascertain. Much as we may dispute with regard to real essences, greatly as we may obscure and confuse our ideas by the attempt to prove that mind is but the result of organized matter, or that matter is but a mode of universal mind, we never in reality doubt that the external world has a real existence. The senses, therefore, are capable of furnishing us with positive knowledge.

220. Prior, in all probability, to the reception of any external impression, there is an internal consciousness of being. It would seem almost necessary that the infant should know something of itself before it begins to perceive anything exterior to itself,—at least the power of perceiving must have an existence (and how can it exist without being known) before perception can take place. But waiving the question of priority as of comparatively little consequence, there are certainly facts of consciousness entirely independent of sensation.

221. We have appetites, passions, desires, sentiments, clearly defined and readily distinguishable from each other. In examining them, we feel that we are examining ourselves; that they are portions of ourselves,—different phases, as it were, of the same indivisible being. Do we ever doubt their reality? Do we ever feel that we may be mistaken in believing that we love or hate, that we fear or venerate, that we hope or despair? You answer, No, emphatically and without hesitation. We know that our own consciousness can never deceive us, and in that, at least, if nowhere else, may we find a sufficient refutation of the skeptical belief that we can be sure of nothing. We know that circumstances cannot affect the reality of our perceptions, that they neither weaken them nor exert any control over them. We find, then, in self-consciousness, a second source of positive knowledge.

222. There is still a third source in the apperceptions of reason. We have a faculty that furnishes us with abstract ideas,—ideas neither of sense nor of consciousness, although they may be first suggested in connection with other truths as the necessary condition of those truths. For example, from sense we derive the idea of body, and in connection with the idea of body, reason at once suggests the necessary and absolute idea of space. Sense discovers the finite, reason mounts to the infinite. Sense perceives the succession of phenomena, reason teaches the relation of cause and effect.

223. These ideas are all essentially distinct, and cannot be confounded with each other. We know that body could not exist without space; the finite without the infinite; the phenomenon without a cause; but we can easily conceive of space without body; of the infinite without any finite existence; of an efficient cause which has the power in itself of either manifesting its efficiency in action, or of remaining entirely at rest.

224. We know, also, that rational ideas cannot be derived from sense or from self-consciousness. We can neither see, nor hear, nor feel, nor smell, nor taste space or infinity or cause; nor can we conceive of them as parts of ourselves. And yet we feel and know that such ideas are types of actual and necessary existence,—that they represent important truths.

225. Besides the power of teaching abstract truth, and the kindred power of generalization, reason perceives the necessary relations of different truths, and is capable of leading us from the simple to the intricate,—from the clear to the more obscure. These relations, when they are plainly perceived, are seen to be unalterable, and founded in the necessity of things. Whatever may be the subject of our consideration, we proceed irresistibly from one conclusion to another, and we feel that from the same data every other rational being would have drawn the same inferences. The propositions of Geometry furnish the most evident proof of this fact, though the proof may be also found in every in-

stance in which one truth is perceived as necessarily resulting from the relation of two or more truths which were previously known.

226. Behold the three guides to knowledge,—the only three that we can possibly employ,—the three within whose province lies the whole territory of conceivable or possible truth. Sense, the guide to a knowledge of the outward world; Self-consciousness, the observer of the inward workings of our own minds; Reason, the teacher of abstract and general truth, and the judge to whose tribunal is our ultimate appeal in all questions of doubt. Distinct, and yet working in entire harmony with each other, they have each a separate and equally important office; the decisions of each in its appropriate sphere are equally reliable. We have seen that this is true in the few instances which have been adduced; and if we extend our inquiries faithfully and cautiously, we shall find that it is always so, and that even the errors to which we are all confessedly subject in no wise weaken the confidence that we naturally repose in each of our faculties.*

227. We are now able to answer our original question: What are the limits and characteristics of positive knowledge? We are necessarily limited to such simple and self-evident propositions as we may be able to discover, and such additional, but more obscure truths as we may logically deduce from a comparison of those elementary propositions. We find in ourselves a tribunal capable of judging in all cases, and if its decisions are pronounced without any hesitation, if they are clearly perceived and understood, and if we feel that they are such as cannot be doubted, we know them to be true.

228. Many of the propositions that receive our full belief are not such as the reason decides upon at once, but their validity is found to rest upon the validity of certain other discoverable data. In examining them, we are obliged to reverse the process by which they were originally acquired, pausing at every step to discover whether Reason gives us her full and unqualified approval. If we can proceed in this manner until we arrive at simple, self-evident propositions, we know that the original propositions are true. Thus, both by deduction from simple truths, and by a critical examination of credible asser-

* In this exposition of the sources of positive knowledge, I have followed pretty closely the teachings of Cousin and Hamilton, introducing such modifications as would give greater prominence to the triple movement of Intelligence under relation. Mahan traces all knowledge to Sense, Consciousness, and Reason, but his definitions appear to limit Consciousness to the sphere of Hamilton's Self-consciousness, and Reason to the sphere of simple Intuition. Self-consciousness is evidently possible only in and through Memory, and if the discussion were merely about the faculties, instead of their modes of action, it would have been more appropriate to have regarded as the three guides to knowledge, the rational faculties through which the first incomes of knowledge are received,—RMM, RMS, RMR. As the symbolism approaches perfection, and its various applications are more fully understood, it will perhaps be easy to define acts and processes with as great precision as we can now define faculties, but for the present, we must content ourselves with such approximations to accuracy as are within our reach.

tions which constitute a portion of our own faith or of the general faith of our race, we are able to enlarge our sphere of knowledge, to replace probability by certainty, and determine the truth or falsehood of much that is involved in doubt.

229. Entire assurance, then, is attainable upon many points, but it can only be attained as the result of patient labor properly directed. We accordingly find a great difference in the precision of ideas, the amount of knowledge, and the degree of confidence in points of belief in different persons. This difference is discernible even in the axioms of different periods, different nations, different individuals, and even of the same individual at different stages of his life.

230. We know that even the axioms of mathematics were not all recognized at once, but they embody the results of long ages of patient investigation. We find that some tribes of men are so wholly unaccustomed to mental discipline, that they will not admit some of the simplest and most evident truths, because they do not understand them; and, finally, in our own experience, we find that as our mind enlarges, not only does our sphere of knowledge enlarge, but we are constantly discovering new simple elementary truths; and even the simpler propositions that once required proof in order to entitle them to our confidence become gradually axiomatic.

231. But with all our progress, whatever may be the character of the age in which we live, or of the circumstances by which we are surrounded, the test of certainty remains the same. We still feel that it rests with us to decide what we know, and what is still unknown, and without inquiring whence we derive the authority to make that decision, we know that we can appeal to none higher or more infallible.

232. "That we cannot show forth *how* the mind is capable of knowing something different from self, is no reason to doubt *that* it is so capable. Every *how* (διότι) rests ultimately on a *that* (ὅτι); every demonstration is deduced from something *given* and *indemonstrable;* all that is comprehensible hangs from some *revealed fact* which we *must believe as actual*, but cannot construe to the reflective *intellect in its possibility.**

233. "The truths known by intuition are the original premises from which all others are inferred. Our assent to the conclusion being grounded upon the truth of the premises, we never could arrive at any knowledge by reasoning, unless something could be known antecedently to all reasoning."†

234. "When men are asked, if any one questions them skilfully, they say all things of themselves, although if they had not an internal knowledge and true reason, they could not do so."‡

* Hamilton, *Discussions*, p. 68. See also Aristotle, 'Αναλυτικῶν ὑστέρων, B. 1, chap. 2, 3.

† J. S. Mill, p. 3.

‡ Plato, *Phædo*, 73 a.

235. Cudworth says: "No man ever was or can be deceived in taking that for a truth which he clearly and distinctly apprehends, but only in assenting to things not clearly apprehended by him." The *probability* of truth may, therefore, reasonably be supposed to be in proportion to the clearness and distinctness of apprehension.

236. Knowledge can extend only so far as our ideas are *clear* and *distinct.** Faith also embraces truths dimly perceived.† The dim perceptions of Faith, in which are included all the mysteries of religion, cannot be made the groundwork or the fit subject of reasoning.‡ In our reverent approaches towards the highest mysteries of Being, in our endeavors to ascertain our relations to the Infinite Loving Father, it soon becomes evident that there are truths far above our mortal ken,§ and we are compelled to satisfy our longing with such dim glimpses and partial disclosures as may be vouchsafed to us individually, or as we may find recorded in the undoubted chronicles of Divine Revelation. The authenticity of a professed revelation is a proper subject for investigation, but after the authenticity is admitted, Reason can deny the teachings of Faith only by abdicating her own throne, which is upheld by other teachings of the same Faith.

237. No one is ever disposed to question the implicit reliance of the child on the instructions of the parent or teacher in whom he has all confidence. Why should we deny

* For some excellent historical and critical remarks on clear, distinct, and confused concepts, see *Hamilton's Logic*, Lect. IX, X.

† "The evidence of things not seen."

‡ We may, it is true, properly speak of a "rational faith," not, however, to imply that its tenets can be either proved or disproved by reason, but merely to indicate that we have sufficient reason for holding the faith. For example, reason may be satisfied that the Bible is the infallible record of Divine Revelation, and belief in the teachings of the Bible then becomes rational faith. But after admitting the plenary inspiration of the Scriptures, reason has no right to sit in judgment over any of their teachings.

"If there is sufficient evidence on other grounds, to show that the Scripture, in which [any] doctrine is contained, is a revelation from God, the doctrine itself must be unconditionally received, not as reasonable, nor as unreasonable, but as scriptural. If there is not such evidence, the doctrine itself will lack its proper support; but the reason which rejects it, is utterly incompetent to substitute any other representation in its place." *Mansel*, p. 168.

§ "We are thus taught the salutary lesson, that the capacity of thought is not to be constituted into the measure of existence; and are warned from recognizing the domain of our knowledge as necessarily coextensive with the horizon of our faith. And by a wonderful revelation we are thus, in the very consciousness of our inability to conceive aught above the relative and finite, inspired with a belief in the existence of something unconditioned beyond the sphere of all comprehensible reality. . . True, therefore, are the declarations of a pious philosophy: A God understood would be no God at all;'—'To think that God is, as we can think Him to be, is blasphemy.' The Divinity, in a certain sense, is revealed; in a certain sense is concealed: He is at once known and unknown." *Hamilton, Discussions*, p. 22. "Canst thou by searching find out God? canst thou find out the Almighty unto perfection?" *Job* 11 : 7.

to the older child the privilege of a like reliance on spiritual guides, who, by greater purity of life, quicker perception of religious truth, consciences more sensitive to the pointings of duty, and continual aspirations for a "closer walk with God," have obtained a degree of religious experience that the mass of mankind, absorbed in the struggles of daily care and toil, could never hope to reach?

238. If I cannot follow all the steps of a Laplace or a Leverrier, I may at least be allowed to rejoice in the faith that the glorious results of their calculations have opened new fields for knowledge and for future investigation. If I cannot understand all the teachings of Paul and John, I have a right to assume, on sufficient evidence, that their lives were purer and their spiritual insight keener than my own; and discarding all hope of attaining to their clearness of vision, I may feel thankful even for the dim perceptions of heavenly glory for which I am indebted to their guidance.

239. The true philosopher should guard carefully against everything like conceit and dogmatism. And especially should those who claim a charitable indulgence for their own conscientious convictions be ever ready to extend a like charity to others.

240. Every one will admit that truth, whether intuitive or demonstrative, revealed or discovered, is always consistent and harmonious, but fancied inconsistency and the consequent fancied fallacy will often prove to be merely imaginary. Because a man's belief, *as I understand it*, appears to be at variance with some fundamental principles of truth, I have no right to pronounce it false as it is held in his own mind. On the contrary, my sense of the entire consistency of all my own personal views should teach me that, although his sphere of vision may be either broader or narrower than my own, he may be compensated by a clearer perception of some points that are shrouded from my eyes in an impenetrable mystery, for his oversight of other points that are to me self-evident. If I charge him with absurdity or folly for professing to believe what is contrary to reason, the charge may recoil on my own head, for it may prove that the folly is mine in assuming my version of his creed to be the correct one. That I should misunderstand him is neither remarkable nor improbable, for even the record of a Divine Revelation, that is worded with all the precision of which human speech is capable, is variously interpreted according to the educational prejudices or mental temperaments of its different readers.

241. The realms of faith and reason, though both harmonious,* are both distinct, and must ever remain so, while we remain less than perfect, and mere learners in the Book of the Universe. If at any future period, during the infinite cycles of eternity, our minds become so thoroughly enlightened that we can blend the two, it must be in the light of a

* Hamilton, *Discussions*, pp. 69, 91, has some very clever remarks on the harmony of truth.

still higher FAITH, that needs none of the slow, hesitating, successive steps of demonstration, but discerns all truth with immediate and intuitive certainty.

242. Reason can only assure us that if the Faith be true on which our premises rest, the conclusion must also be true, but of the probability or certainty of the primary beliefs which constitute the materials of our reasoning, Faith itself is the only judge.*. We may, however, safely rest in the assurance that the merciful Providence of an All-wise Creator will always vouchsafe to each of his intelligent creatures all the revelations that are essential to his continual needs; and that if we act at each moment in accordance with the light that is given us, we shall not only perform the duty of the moment, but we shall also make continual spiritual progress,—attaining to a clearer understanding of truths that we have already learned, and gradually extending the sphere of our belief until it embraces every vital doctrine of revelation.

CHAPTER VII.

RATIONAL ANTINOMIES.

243. WHILE apparent antagonisms of Faith are, therefore, perfectly legitimate, there can be no legitimate antagonisms of Reason, either real or apparent.

244. It is true that the ancient philosophers often puzzled and amused themselves with paradoxes, dilemmas, and paralogisms or sophisms, but even when they were unable to detect the fallacy, they always felt that it must arise from some unwarranted use of terms. No one appears to have taught that Reason could be rightfully involved in necessary and irreconcilable opposition, until Kant propounded his celebrated Antinomies.

245. He says: "If we apply our Reason, not merely for the use of the principles of the understanding to objects of experience, but venture to extend such out beyond the limits of the latter,† sophistical theorems thence arise, which neither need look for confirmation in experience, nor fear opposition, and each of which is not only in itself without contradiction, but in fact finds, in the nature of reason, conditions of its necessity, only that, un-

* "The whole province of faith belongs *objectively* to reason too; for if faith made us believe what is unreasonable *in itself*, it would be an unreasonable, and therefore a false faith, and one we should be better without. Faith is but the advanced guard, marching onward through the territory really belonging to Reason, though not actually occupied by it; and the broader the base of operations covered by Reason, the farther may Faith itself advance, without danger of stumbling upon the outposts of error." *Solly*, p. 16.

† In other words, if we try to reason upon subjects that are beyond our power of comprehension.

fortunately, the contrary has equally as valid, and as necessary grounds of affirmation on its side."*

246. All legitimate reasoning, as we have seen, requires that the premises, as well as their relations, should be clearly apprehended. If we transcend the limits of possible experience, clear apprehension becomes impossible, and as we have no means of fixing and defining our ideas, we are easily led into confusion and equivocation.

247. The four Kantian antinomies furnish admirable illustrations of this truth. They all relate to different ideas of the ABSOLUTE: 1. The Absolute completeness of the composition of the given whole of all phenomena; 2. The Absolute completeness of the division of a given whole in the phenomenon; 3. The Absolute completeness of the arising of a phenomenon in general; 4. The Absolute completeness of the dependency of the existence of the changeable in the phenomenon.†

248. At the very outset, we are confused by the vagueness of the term Absolute. It "is of a twofold (if not threefold) ambiguity, corresponding to the double (or treble) signification of the word in Latin.

"1. *Absolutum* means what is *freed* or *loosed*, in which sense the Absolute will be what is aloof from relation, comparison, limitation, condition, dependence, &c., and thus is tantamount to τὸ ἀπόλυτον of the lower Greeks. In this meaning the Absolute is not opposed to the Infinite.

"2. *Absolutum* means *finished, perfected, completed*; in which sense the Absolute will be what is out of relation, &c., as finished, perfect, complete, total, and thus corresponds to τὸ ὅλον and τὸ τέλειον of Aristotle. In this acceptation,—and it is that in which for myself I exclusively use it,—the Absolute is diametrically opposed to, is contradictory of, the Infinite.

"Besides these two meanings, there is to be noticed the use of the word, for the most part in its adverbial form;—*absolutely* (*absolute*) in the sense of *simply, simpliciter* (ἁπλῶς), that is, considered in and for itself,—considered not in relation."‡

249. The philosophical Absolute is nearly always,—perhaps always,—considered as unlimited, but there is a great difference of opinion as to what constitutes a limit. Some regard any affirmation or negation as a limit,—others regard that as finite which has any qualities or attributes, and they approach the absolute by excluding all attributes. The highest form of simple attribution, is generally admitted to be that of Existence or Being. If from Being we suppose the attribute of Being to be excluded, we may call the supposed resultant the Absolute. But if this Absolute is anything that we can think about,—inas-

* P. 299.

† Kant, p. 295.

‡ Hamilton, *Discussions*, p. 21.

much as thought implies attribution,—we must still exclude the attribute of Absoluteness, which brings us to the Zero, or Pure Nothing of Hegel and Oken.*

250. It seems strange that any one should ever have attempted to reason about an idea that is so vague, indefinite, and indefinable, and yet on such groundless reasoning have been mooted some of the most profound problems of metaphysics. The arguments, of course, are all drawn from that which is supposed to be known,—from the finite and relative,—and as it is easy to find opposing and contradictory relations, we may easily obtain contradictory conclusions, provided we take the first false step of admitting that relation, or the consequences of relation, can be predicated of that which is devoid of all relation.

251. Kant disposes of the Antinomies properly enough, by what he terms the *Skeptical method*, that is to say, by inquiring whether the object of dispute "may not, perhaps, be a mere delusion, at which each catches in vain, and whereby he can gain nothing, although he were not at all to be opposed."† In some cases, however (*e. g.*, in the mathematical fallacies, pp. 506–7), the skeptical method would not be applicable, not, as Kant states, because "its use would be absurd,"† but because every false, absurd, or contradictory conclusion results from the employment of equivocal premises, and the philosophical investigator should endeavor to trace the equivocation to its lurking-place.

252. "THE ANTINOMY OF PURE REASON.‡

"First Contradiction of Transcendental Ideas.

"THESIS.

"The world has a beginning in time, and is also inclosed as to space, in limits.

"Proof.

"For, if we admit that the world has no commencement as to time, an eternity, then, has elapsed up to each given point of time, and consequently, an infinite series of states of things, following upon one another in the world, has passed away. But now the infinity of a series consists in this very thing,—that it can never be completed by successive synthesis. Consequently

"ANTITHESIS.

"The world has no beginning, and no limits in space, but is, as well in respect of time as of space, infinite.

"Proof.

"Let it then be supposed that it has a beginning. As the Beginning is an existence which a time preceded, wherein the thing is not; a time must have gone before, wherein the world was not, that is, a void time. But now, in a void time, no origin of anything is possible, because no part of such a time has in itself, prior to another, any distinctive condition of ex-

* "The Intuition of God = the Absolute = the Nothing, we [also] find asserted by the lower Platonists, by the Buddhists, and by Jacob Boehme." Hamilton, *Discussions*, p. 28.

† P. 301.

‡ Kant, pp. 303–307.

an infinite elapsed cosmological series is impossible, therefore a beginning of the world is a necessary condition of its existence, which first was to be shown.

" In respect to the second point, if we again maintain the contrary, the world will thus be an infinite given whole of contemporaneously existing things. Now we cannot think the magnitude of a Quantum,* which is not given within certain limits of every intuition, in any other way than through the synthesis of the parts, and the totality of such a Quantum, only through the completed synthesis, or through repeated addition of unity to itself.† Hence, in order to think the world, which fills all space as a Whole, the successive synthesis of the parts of an infinite world must be looked upon as completed, that is, an infinite time must be looked upon as elapsed in the enumeration of all coexistent things; which is impossible. Consequently, an infinite aggregate of real things, cannot be looked upon as a given whole, and therefore not as given *contemporaneously*. Thus a world is *not* in respect of its extension in space *infinite*, but inclosed in its limits; which was the second point.

* "We can envisage an undetermined Quantum as a whole, if it is inclosed in limits, without requiring to construct the totality thereof by measurement, that is, the successive synthesis of its parts. For the limits determine already the completeness, since they cut off all moreness.

† "The conception of totality is, in this case, nothing else but the representation of the completed synthesis of its parts, since as we cannot deduce the conception from the intuition of the whole (which in this case is impossible), we can only comprehend this whole by means of the synthesis of the parts, up to the completion of the infinite, at least in idea.

istence rather than of non-existence (whether we admit that this condition arises of itself, or through another cause). Several series of things can, therefore, indeed begin in the world, but the world itself can have no beginning, and therefore, is in respect of elapsed time, infinite.

" As to what concerns the second point, let us first take the contrary, that is to say, that the world in respect of space, is finite and limited; it finds itself, in this way, in a void space, which is not limited. There would, therefore, be met with, not only a relationship of things in *space*, but also of things *to space*. Now as the world is an absolute Whole, without of which no object of intuition, and consequently no correlative of the World is found, wherewith the same stands in relationship, the relationship of the World to void space would thus be a relationship thereof to *no object*. But such a relationship, and therefore, the limitation of the World by void space is nothing; consequently, the World in respect of Space is not at all limited, that is to say, in regard to extension it is infinite.*

* "Space is merely the form of the external intuition (formal intuition), but no real object that externally can be envisaged. Space before all things which determine it (fill or limit), or rather which afford an *empirical intuition* according to its form, is under the name of absolute space, nothing else but the mere possibility of external phenomena, so far as they either exist of themselves, or can yet be added to given phenomena. The empirical intuition is, therefore, not composed of phenomena and space (perception and void intuition). One is not correlative of the synthesis of the other, but only conjoined in one and the same empirical intuition, as matter and form thereof. If we will place one of these two points out of the other (space out of all phenomena), there arises thence all kind of void determinations of the external intuition, which still are not possible perceptions. For example, motion or rest of the world in infinite void space, a determination of the relationship of the two with one another, which never can be perceived, and is, therefore, likewise the predicate of a mere ideal thing.

"OBSERVATION UPON THE FIRST ANTINOMY.

"1. *Upon the Thesis.*

"I have not sought after deceptions in these mutually contradictory arguments in order, for instance (as it is termed), to advance an advocate's proof, who avails himself of the imprudence of his opponent for his own advantage, and willingly sanctions his appeal to a misunderstood law in order to establish his own unjust pretensions upon the refutation of it. Each of these proofs is deduced from the nature of things, and the advantage set aside which the erroneous conclusions of Dogmatists could afford us on both parts.

"I might, likewise, have been able to demonstrate according to appearance the Thesis, by reason of this, that I premised, agreeably to the custom of the Dogmatists, an erroneous conception as to the infinity of a given quantity. A quantity is infinite, beyond which no greater (that is, beyond the therein contained multiplicity of a given unity) is possible. Now, no multiplicity is the greatest, inasmuch as always one or more unities can still be added thereto. Consequently an infinite given quantity,—consequently, also (in respect of the elapsed series as well as of extension), an infinite world is impossible. It is, therefore, in both ways limited. I might, in such a way, have adduced my proof; but this conception does not accord with that which we understand by an infinite whole. It is not, thereby, represented so great as it is; consequently, also, its conception is not the conception of a *maximum*, but only, thereby, its relationship to an arbitrarily to be adopted unity is thought, in respect of which this relationship is greater than all number. Now, accordingly as unity is admitted greater or less, the infinite would be greater or less; but infinity, as it consists merely in the relationship to this given unity, would remain ever the same, although certainly the absolute quantity of the whole thereby would not be at all known—but as to which it is not here the question.

"The true (transcendental) conception of infinity is that the successive synthesis of unity in the measure-

"OBSERVATION.

"2. *Upon the Antithesis.*

"The proof of the infinity of the given cosmological series, and of the cosmological Whole, rests upon this: that in the opposite case a void time as well as a void space must constitute the limits of the world. Now, I am not ignorant that against this consequence excuses are sought for, inasmuch as it is pretended that there is a limit of the world in respect of time and space quite possible, without its being even requisite to admit an absolute time before the beginning of the world, or an absolute extended space out of the real world, which is impossible. I am entirely satisfied with the last part of this opinion of the philosophers of the Leibnitzian school. Space is merely the form of the external intuition, but no real object which can be envisaged externally, and no correlative of phenomena, but the form of phenomena themselves. Space, therefore, cannot absolutely (of itself alone) occur as something determining in the existence of things, since it is no object at all, but only the form of possible objects. Things, therefore, as phenomena, certainly determine space; that is, under all possible predicates thereof (quantity and relationship), they so operate that these or those belong to reality; but conversely, space, as something which subsists of itself, cannot determine the reality of things in respect of the quantity or form, because in itself it is nothing real. Consequently, a space (whether full or void)* may very well be limited by phenomena, but phenomena can never be *limited* by *means of a void space* external to them. The same is also valid as to time. But all this being granted, it is still, nevertheless, indubitable that we must absolutely admit two nonentities, void space out of the world, and void time before

* "It is easy to be observed, that hereby it is intended to say, that *void space so far as it is limited by phenomena*—consequently that such *within the world* does not, at least, contradict the transcendental principles, and may, therefore, be admitted in respect of the same (although its probability is not, on that account, directly maintained).

ment of a Quantum can never be completed.* Hence, it follows quite certainly that an eternity of real states following upon one another can never have elapsed up to a given (the present) point of time,—consequently, the world must have a beginning.

"In regard to the second part of the thesis, the difficulty certainly disappears of an infinite and yet elapsed series, for the diversity of an infinite world, as to extension, is given *coexistently*. But in order to think the Totality of such a multiplicity, since we cannot appeal to limits which constitute the totality in itself in the intuition, we must render an account of our conception, which, in such a case, cannot go from the whole to the determined multiplicity of the parts, but must show the possibility of a whole by means of the successive synthesis of the parts. And as this synthesis must form a never to be completed series, we cannot thus think a totality prior to it (*the synthesis*), and consequently, also, not through it. For this conception of totality itself is, in this case, the representation of a completed synthesis of parts, and this completion, and consequently the conception thereof, is impossible."

* "This (the Quantum) thereby contains a multiplicity (of given unity), which is greater than all number, which is the mathematical conception of the infinite."

the world, provided we admit a limit to the world whether in respect of space or time.

"For as to what regards the subterfuge whereby we strive to avoid the consequence, agreeably to which we say that if the world (according to time and space) has limits, the infinite void must determine the existence of real things in respect of their quantity; it consists thus only in this, that we think to ourselves instead of a *sensible world*, some sort of an intellectual world, and instead of a first beginning (an existence previous to which a time of non-being precedes), an existence generally is imagined, which *presupposes no other condition* in the world, and instead of boundaries of extension, limits are conceived of the universe, and thereby avoidance is made of time and space. But here the question is only as to *mundus phænomenon* and its quantity, in respect of which we can, by no means, make abstraction of the stated conditions of sensibility without annihilating the being of it. The sensible world, if it be limited, lies necessarily in the infinite void. If we will omit this, and consequently space in general as condition of the possibility of phenomena *à priori*, the whole sensible world then disappears. In our problem this alone is given us. The *mundus intelligibilis* is nothing but the universal conception of a world in general, in which conception we make abstraction of all conditions of the intuition of this world, and in respect of this conception, no synthetic proposition, either affirmative or negative, is possible."

CHAPTER VIII.

EXAMINATION OF ANTINOMIES.

253. THE first ambiguity that presents itself in the foregoing antinomy, is in the meaning of the term "world." Kant says, "The ideas with which we now concern ourselves I have before termed Cosmological ideas, partly on this account, because under world the complex of all phenomena is understood, and our ideas also are only directed to the un-

conditioned amongst phenomena;* partly, likewise, because the word world, in a transcendental sense, signifies the absolute totality of the complex of existing things, and we direct our attention alone to the completeness of the synthesis (although only strictly in the regressus to the conditions)."† The arguments of the Thesis are mainly based on ideas derived from the first of these definitions, and they have a quasi validity to show that the aggregate of phenomena (if the cause of the phenomena is excluded from consideration) may have had a beginning in time and a limit in space. The Antithesis can only be valid for the second definition to show that "the absolute totality of the complex of existing things" (including the Creative First Cause together with every possible and actual manifestation of His existence and power) must be, "as well in respect of time as of space, infinite."

254. The second ambiguity is in the use of the word "infinite." In the sense in which some philosophers have employed the terms infinite and finite, they are mutually contradictory, and it is as absurd to speak of their correlation as it would be to talk of the four sides and six angles of a square triangle. In one sense, the mere formation of an idea is limiting, inasmuch as it assigns bounds which distinguish the idea from all others, but according to customary usage, we have a right to call that infinite which is unlimited in one or more of its attributes.‡ We have no right, however, to assume that what is true of one infinite is true of another, as is repeatedly done in each of the Kantian antinomies.

255. In the celebrated sophism of Achilles and the tortoise,§ there is a similar equivocal use of the ideas of infinity. "The fallacy, as Hobbes hinted, lies in the tacit assumption that whatever is infinitely divisible is infinite. . . . The 'forever' in the conclusion means for any length of time that can be supposed; but in the premises 'ever' does not mean any *length* of time; it means any *number of subdivisions* of time. It means that we may divide a thousand feet by ten, and that quotient again by ten, and so on as

* But an unconditioned phenomenon is an impossibility.

† P. 298.

‡ Werenfels, *De Finibus Mundi Dialogus* (quoted by Mansel, p. 253), ingeniously attempts to demonstrate that the idea of infinite extension involves necessary contradictions. But the whole argument is based on the unwarranted assumption that all relative infinites are equal. It is important, even in discoursing on ordinary topics, that all the conditions of the several points at issue should be kept in view, and this precaution is still more essential in reasoning upon a subject so obscure as infinity.

§ "Let Achilles run ten times as fast as the tortoise, yet if the tortoise has the start, Achilles will never overtake him. For suppose them to be at first separated by an interval of a thousand feet, when Achilles has run these thousand feet, the tortoise will have got on a hundred; when Achilles has run those hundred the tortoise will have run ten, and so on forever; therefore, Achilles may run forever without overtaking the tortoise." See *Mill's Logic*, p. 508, and Aristotle, φυσιχῆς ἀκροάσεως, B. vi, chap. 9, p. 549.

often as we please; that there never needs be an end to the subdivisions of the distance, nor, consequently, to those of the time in which it is performed. But an unlimited number of subdivisions may be made of that which is itself limited. The argument proves no other infinity of duration than may be embraced within five minutes. As long as the five minutes are not expired, what remains of them may be divided by ten, and again by ten, as often as we like, which is perfectly compatible with their being only five minutes altogether. It proves, in short, that to pass through this finite space requires a time which is infinitely divisible, but not an infinite time."*

256. Kant's definition, that "the Infinity of a series consists in this very thing, that it can never be completed by successive synthesis," should be qualified by adding,—*unless the succession is infinite.* Every finite quantity, being infinitely divisible, is the completion or sum of an infinite number of infinite series; every *now* is the termination of one infinity, and the commencement of another infinity of successive moments. It is quite true, as Kant remarks in his Observation upon the Thesis, "that an eternity of real states following upon one another, can never have elapsed up to a given (the present) point of time," provided we mean by eternity, duration that has neither beginning nor end, but our Thesis and Antithesis refer merely to the beginning, and it is quite certain that a terminated "succession of real states," infinite in regard to its commencement, must have elapsed at every given point of time.

257. A third ambiguity arises from the equivocal meaning of *space* and *time.* In a portion of the reasoning, they are regarded as mere forms of thought; in another portion, as real entities. If space is included in the phenomena of the world, being itself infinite, the world must also be infinite. But if space is a mere form of thought, and in no sense phenomenal, we may easily imagine "not only a relationship of things in *space*, but also of things *to space*," and the world may, therefore, be conceived as "inclosed as to space, in limits."

258. This brief discussion is, perhaps, sufficient to show that the Antinomies do not necessarily result from the legitimate use of Reason, but that they are pure fallacies, and that, like other fallacies, they will be self-detected, provided all the terms are clearly and unequivocally defined.

259. The following are the remaining Kantian Antinomies:†

* Mill, *Logic*, p. 508. Mill disclaims the invention of this solution, but does not mention the author. I thought it was from Hamilton, but I have not been able to turn to it.

† Kant, pp. 308, 314, 319.

II.

"THESIS.

"Every compound substance in the world consists of simple parts, and there exists everywhere nothing but the simple, or that which is compounded from it.

"ANTITHESIS.

"No compound thing in the world consists of simple parts, and there exists nothing anywhere therein simple.

III.

"Causality, according to the laws of nature, is not the only one from which all the phenomena of the world can be derived. There is, besides, a Causality through liberty, necessary to be admitted for the explanation of the same.

"There is no liberty, but everything in the world occurs only according to laws of nature.

IV.

"Something belongs to the sensible world, which either as its part, or its cause, is an absolutely necessary being."

"There exists nowhere any absolutely necessary being, neither in the world nor out of the world, as its cause."

260. The principal ambiguities in these several Antinomies are the following:

II. A compound may be either chemical, of things differing in sensible qualities,—mathematical, of things differing in position,—or immaterial, of things differing in ideal relations.

III. Absolute or unlimited liberty is inconceivable, but a liberty within certain limits may be subject to laws of its own, of which the laws of nature may be considered either as inclusive, or as exclusive.*

IV. The necessary First Cause may either be considered separately from the aggregate of phenomena, or it may be regarded as a portion of "the absolute totality of the complex of existing things."

261. Hamilton gives some "contradictions proving the psychological theory of the conditioned,"† which look strangely out of place in a work by a disciple of the Scotch school, but which result necessarily from his questionable qualifications of the theory, "that all

* I cannot imagine any more concise and satisfactory solution of Kant's third Antinomy (provided we assume that the unconditioned is a fit subject for reasoning), than the one given by Solly (p. 24–5). "Now the unconditioned cause is necessarily free; for were it not so, it would be subject to a condition, a supposition which is excluded by the hypothesis. The conditioned cause, on the other hand, is necessarily not free, for otherwise it would not be limited by a condition, which is equally excluded by the hypothesis. If, however, we take the whole of nature, and seek for its cause, inasmuch as it comprises all conditioned causes within itself, the cause in question must clearly be unconditioned and free. While therefore the causality *in* nature is conditioned, the cause *of* nature itself is unconditioned." [Should we not rather say, the cause of nature is only self-conditioned?]

† *Metaphysics*, p. 682. See also pp. 527-31.

that is conceivable in thought, lies between two extremes, which, as contradictory of each other, cannot both be true, but of which, as mutual contradictories, one must." The most important of these supposed contradictions deserve a passing notice.

"1. Finite cannot comprehend, contain the Infinite.—Yet an inch or minute, say, are finites, and are divisible *ad infinitum*, that is, their terminated division is incogitable.

"2. Infinite cannot be terminated or begun.—Yet eternity *ab ante* ends *now*; and eternity *a post* begins *now*. So apply to Space.

"3. There cannot be two infinite maxima.—Yet eternity *ab ante* and *a post* are two infinite maxima of time.

"4. Infinite maximum if cut into two, the halves cannot each be infinite, for nothing can be greater than infinite, and thus they could not be parts; nor finite, for thus two finite halves would make an infinite whole.

"5. What contains infinite quantities, cannot be passed through,—come to an end. An inch, a minute, a degree contains these; *ergo*, etc. Take a minute; this contains an infinitude of protended quantities, which must follow one after another; but an infinite series of successive protensions can, *ex termino*, never be ended; *ergo*, etc.

"6. An infinite maximum cannot but be all inclusive. Time *ab ante* and *a post* infinite and exclusive of each other; *ergo*.

"7. An infinite number of quantities must make up either an infinite or a finite whole. I. The former.—But an inch, a minute, a degree, contain each an infinite number of quantities; therefore, an inch, a minute, a degree, are each infinite wholes; which is absurd. II. The latter.—An infinite number of quantities would thus make up a finite quantity; which is equally absurd.

"8. If we take a finite quantity (as an inch, a minute, a degree), it would appear equally that there are, and that there are not, an equal number of quantities between these and a greatest, and between these and a least. . . .

"13. A quantity, say a foot, has an infinity of parts. Any part of this quantity, say an inch, has also an infinity. But one infinity is not larger than another; therefore an inch is equal to a foot."*

262. The ambiguity that runs through all these propositions, is the same that has already been noticed. In each proposition, the term infinite is used with two or more meanings, and the different properties of different relative infinites, are contrasted with the supposed properties of a supposed absolute infinite. Hamilton himself points out this

* This fallacy resembles the algebraical demonstration already given, that $1 = 3$. If one infinity is not larger than another, then $\frac{1}{0} = \frac{n}{0}$, and $1 = n$. In reality, o and ∞ may each have an infinite number of values, and any reasoning that is based either upon the infinitely great or the infinitely small, may lead us into error, unless we keep all the conditions in view,—those which are limiting, as well as those which are infinite.

ambiguity in his letter to Mr. Henry Calderwood, in which he remarks, "that there is a fundamental difference between *The Infinite* (τὸ ἓν καὶ πᾶν), and a relation to which we may apply the term *infinite*."* We can reason correctly about the relatively infinite, but not about the absolutely infinite, which is devoid of all relation, even of the relations of unity, reality, and conceivability.†

263. Hamilton evidently uses the term conceivable in a narrower sense than many other philosophical writers, and I am inclined to believe that if his meaning were made perfectly clear, the truth of his Law of the Conditioned, "that the conceivable is in every relation bounded by the inconceivable," would be generally admitted. In the letter which has just been quoted, he says, "What I have said as to the infinite being (subjectively) inconceivable, does not at all derogate from our belief of its (objective) reality. In fact, the main scope of my speculation is to show articulately that we *must believe* as actual, much that we are unable (positively) *to conceive*, as even possible."‡ But when he says that "though space must be admitted to be necessarily either finite or infinite, we are able to conceive the possibility neither of its finitude, nor of its infinity,"§ he seems to be struggling with a perplexity that might have been avoided, if his ideas had been more clearly defined.

* *Metaphysics*, p. 685. See also Kant's Observations on the First Antinomy.

† In designating unity, reality, and conceivability as relations, I do not refer to the category of relation, but to the idea of relativity which underlies all human thought, and which Hamilton regarded as conditioning every object of thought. Thus, in what seems a petitio principii, he says (*Discussions*, p. 21), "Thought is only of the conditioned; because as we have said, to think is simply to condition. The *absolute* is conceived merely by a negation of conceivability, and all that we know, is only known as

——'won from the void and formless *infinite*.'

"How, indeed, it could ever be doubted that thought is only of the conditioned, may well be deemed a matter of the profoundest admiration. Thought cannot transcend consciousness; consciousness is only possible under the antithesis of a subject and object of thought, known only in correlation, and mutually limiting each other; while, independently of this, all that we know either of subject or object, either of mind or matter, is only a knowledge in each of the particular, of the plural, of the different, of the modified, of the phenomenal."

How will these remarks apply when the subject and object are one,—the subject thinking of itself? We certainly can think *of* the unconditioned, the absolute, the infinite, even if we are obliged to condition them in our endeavors to understand them. All relative infinites are certainly included in absolute infinites, and in some sense as parts of the absolute. Can the effort to abstract all relativity, and thus arrive at an Absolute or Unconditioned, result in anything else than the Hegelian o, or Hamilton's *Nihil purum?* Can anything exist except in relations, either internal or external? Is there a Unity that embraces Infinite Space, Eternity, Matter, Mind, Possibility, Relation, and Truth, but is yet neither of these, and in no relation to either? Such is certainly not the teaching of revelation, or of any intelligible philosophy. [For some excellent remarks on the relations of the Infinite, see Catherwood, pp. 103 *et seq.*]

‡ *Metaphysics*, p. 687. By a "positive conception," Hamilton evidently means a complete or *adequate* notion.

§ Ibid. p. 527.

264. Mill, in his Chapter on Fallacies of Simple Inspection, very properly controverts the proposition, that whatever is inconceivable must be false. Some of his reasoning, however, is not very conclusive. Take, for example, the following passages.

265. "Rather more than a century and a half ago, it was a philosophical maxim, disputed by no one, and which no one deemed to require any proof, that 'a thing cannot act where it is not.' With this weapon the Cartesians waged a formidable war against the theory of gravitation, which, according to them, involving so obvious an absurdity, must be rejected *in limine;* the sun could not possibly act upon the earth, not being there. It was not surprising that the adherents of the old systems of astronomy should urge this objection against the new; but the false assumption imposed equally upon Newton himself, who in order to turn the edge of the objection, imagined a subtle ether which filled up the space between the sun and the earth, and by its intermediate agency was the proximate cause of the phenomena of gravitation. . . .

266. "No one now feels any difficulty in conceiving gravity to be, as much as any other property is, 'innate, inherent, and essential to matter,' nor finds the comprehension of it facilitated in the smallest degree by the supposition of an ether; nor thinks it at all incredible that the celestial bodies can and do act where they, in actual bodily presence, are not. To us it is not more wonderful that bodies should act upon one another, 'without mutual contact,' than that they should do so when in contact; we are familiar with both these facts, and we find them equally inexplicable, but equally easy to believe. . . .

267. "It is strange that any one, after such a warning, should rely implicitly upon the evidence, *à priori*, of such propositions as these, that matter cannot think; that space, or extension, is infinite; that nothing can be made out of nothing (*ex nihilo nihil fit*). Whether these propositions are true or no, this is not the place to determine, nor even whether the questions are soluble by the human faculties. But such doctrines are no more self-evident truths than the ancient maxim that a thing cannot act where it is not, which probably is not now believed by any educated person in Europe."*

268. This whole course of argument rests so evidently on ambiguity of definition, that it furnishes an admirable exemplification of the origin of all antinomies. No one ever questioned the statement that force can be transmitted from one point to another, and the fact of such transmission does not prove that a thing can act where it is not. In machinery that is worked by steam, the steam acts "where it is," on the piston of the engine, but its force may be conveyed through a series of mechanical means, and finally used at a point very remote from the boiler. An electro-magnetic battery acts "where it is," on the wire at a telegraph-station, but the force that it communicates may be conducted thou-

* Mill's Logic, pp. 461, 462.

sands of miles before it reaches its destination. The earth falls towards the sun in consequence of the attractive force that acts upon the earth "where it is." The belief is almost universal that the force is transmitted from the sun to the earth in some unknown way.

269. There are many reasons for supposing that all the imponderable agents, light, heat, electricity, attraction, are different modifications of FORCE, all acting in similar ways and subject to similar laws, but differing in their effects on account of the different circumstances attending their action. There are no greater difficulties connected with Newton's hypothesis of a subtle ether through which gravitation may be transmitted, than with the hypothesis of a similar ether to sustain the undulatory theory of light. The general reception of the undulatory theory proves that every "educated person in Europe" and elsewhere still feels the necessity of endeavoring to account for all transmission of force, and still believes that "a thing cannot act where it is not" in the sense in which Newton probably believed the maxim, though it is doubtful whether any one ever believed it in the sense that Mill disputes.

270. In like manner the remaining propositions about matter, space, and creation can, undoubtedly, be so defined that their *à priori* self-evidence may be doubted or even denied. But, in what I regard as the common acceptation of the terms, I cannot but think that I have a right positively to assert "that matter cannot think; that space, or extension, is infinite; that nothing can be made out of nothing." Of these several assertions, the last seems the most questionable, but in order that anything may be made, there must at least be a maker who has the power of making. The power of doing anything implies the exertion of force, and whatever is produced by the maker, exerting the force that is in his power, cannot, in every possible sense, be said to be "made out of nothing."

271. If the instances that have been adduced by profound students of philosophy to prove that reason is sometimes entangled in an inexplicable dilemma, are divested of difficulty when all the terms are used with a clear and intelligible meaning, there can be little risk in repeating the assertion that there can be no legitimate antagonisms of Reason either real or apparent. Whenever any line of argument appears to lead to a paradox, it may safely be inferred that Reason has either left her own province or that she has become confused by some hidden equivocation. Every honest critic, therefore, should make due allowance for the imperfections of language and the consequent danger of misapprehension, and if he can discover in the propositions that he is considering any truthful meaning, he should regard that meaning as the one that the author intended to convey.

CHAPTER IX.

BASIS OF ONTOLOGY.

272. In extending our inquiries beyond the mind and its capacities, to the unthinking objective, our conclusions must rest entirely upon faith. We know what takes place in our own minds, and we know how we are affected by external bodies, but what is the condition of those bodies, or how truly that condition is represented by our conceptions, it is impossible for us to know.

273. If the subjective is limited to the sphere of human consciousness, we can judge of the objective-objective relation, or of the objective sides of the objective-subjective and the subjective-objective relations, only by an assumed analogy between our cognition of phenomena and their supposed cause.

274. Our faith in such an analogy may be strengthened by the rational conviction that the highest unity, in which both the knowing and the knowable are joined, must be a Supreme Intelligence;* that the source of all things is therefore an Infinite Omniscient Subjective,—and that, so far as our finite subjective resembles the Infinite, our subjective views of the objective will resemble the higher subjective reality of the objective, as it is perceived by the Infinite Intelligence.

275. Since the days of Wolff, the term Ontology has generally been applied to the science of Being,†—the science of the purely objective. Inasmuch as the realm of Ontology lies entirely outside of all possible experience, its data, like the primary cognitions that furnish the conditions of experience and of reason, transcend the sphere of reason, and, therefore, belong to what has been called, since the days of Kant, Transcendental Philosophy.

276. The reasons that have been given for adopting a trichotomy in the investigation of mental phenomena, are equally valid for a primary fundamental analysis of the objective, that is based upon mental analogy. The whole sphere of Being, above Consciousness, can hardly be well studied in any other way than in its relations to the triform Consciousness; but after reaching the level of Consciousness, the subsequent divisions are within the possible relations of experience, and "the same subject may admit, and even

* "Matter does not move matter otherwise than as a medium, but Mind does move it." *Taylor: World of Mind*, p. 35.

† See Fleming, p. 162.

require, various divisions, according to the different points of view from which we contemplate it; nor does it follow that because one division is good, therefore another is naught."*

277. It may, perhaps, be granted that the ternary division, founded as it is on necessary and immutable relations, would be the most purely philosophical for all Science,—the empirical as well as the transcendental. But it is impossible, as yet, to do more than to lay the broad basis for generalization, and to make an experimental essay with some of the fundamental branches of knowledge.

278. When this essay has been thoroughly tested and fully approved, another step may be taken, and gradual approaches may thus be made to a grand schedule of the knowable, which will furnish, by its symbolism, a universal language that will be as easily read and understood as the symbolic language of Arithmetic and Algebra. Meanwhile, each investigator, pursuing his own specialty in his own way, will be accumulating materials for some department of Universal Science, to be fitly and permanently arranged at some future day, if the dreams of philosophy are ever realized.

279. Even the founder of the modern school of "Positive Philosophy" adopts the trinal basis, but without recognizing the source of the fundamental law that he assumes from observation. He says:

280. "From the study of the development of human intelligence in all directions and through all times, the discovery arises of a great fundamental law to which it is necessarily subject, and which has a solid foundation of proof, both in the facts of our organization and in our historical experience. The law is this: that each of our leading conceptions—each branch of our knowledge—passes successively through three different theoretical conditions: the Theological, or fictitious; the Metaphysical, or abstract; and the Scientific, or positive. . . . Hence arise three philosophies, or general systems of conceptions on the aggregate of phenomena, each of which excludes the others. The first is the necessary point of departure of the human understanding, and the third is its fixed and definite state. The second is merely a state of transition."†

281. It is difficult to understand how any branch of knowledge, that is in its first stage fictitious, can subsequently become abstract and finally positive. If we modify the conditional formula so as to read,—1, the Theological or credible, resting on faith in the irresistible beliefs implanted in us by the Creator; 2, the Metaphysical or abstract, examining the validity of reason in its deductions from faith; 3, the Scientific or positive, embracing all the legitimate teachings of faith and reason;—the gradation will, perhaps,

* Reid, p. 688

† Comte, pp. 25, 26.

be more natural and more logical, as well as in strict accordance with the successive Motive, Spontaneous, and Rational development of Consciousness.

282. Aristotle's division of fundamental science recognizes the supreme importance of theology:

"Physical science is about those things which have in themselves the principle of motion; but mathematical science is contemplative, and about permanent but inseparable things. Therefore, about separable and immovable being, if there is any such being, there is another science besides these two. I say separable and immovable, which we will endeavor to show; and if there is any such nature in beings, there also would be the divine; and this would be the first and supreme principle. It is evident, therefore, that there are three kinds of contemplative sciences, PHYSICAL, MATHEMATICAL, THEOLOGICAL. The class of contemplative sciences is, therefore, the best; and of these, the one last mentioned, for it is about the most honorable of beings."*

283. A Positive Philosophy is possible to those, and only to those, who have a positive faith. In all reasoning, it is necessary to inquire, not only whether the argument is logical, but also whether the premises are true. The theological condition of knowledge is not only the first, but it is the most continuous and the most authoritative.

284. "We do not see a man, if by Man is meant that which lives, moves, perceives, and thinks as we do; but only such a certain collection of ideas as directs us to think there is a distinct principle of thought and motion like to ourselves, accompanying and represented by it. And after the same manner we see GOD; all the difference is that whereas some one finite and narrow assemblage of ideas denotes a particular human mind, whithersoever we direct our view, we do at all times and in all places perceive manifest tokens of the Divinity; everything we see, hear, feel, or anywise perceive by sense, being a sign or effect of the power of GOD; as is our perception of those very motions which are produced by man."†

285. "If mind have no original or existence but with us, by what means or way do we men come to be possessed of it? Our principle of intelligence, or soul, which has dominion over our body, is no more visible to us than the principle of the universe."‡

* Aristotle,—τῶν μετὰ τὰ φυσικά, B. xi, chap. 7, vol. 2, p. 1381. See also B. vi, chap. 1, p. 1308.

† Berkeley: *Treatise concerning the Principles of Human Knowledge*, § 148. Compare Solly, p. 239. "The only alternative, as it appears to me, which saves any form of life and freedom in the external world, anything that should account for and justify the profound sense of awe we experience in viewing the glories of the universe, without making it the very God Himself, and thus rushing at once into the very grossest form of pantheism, is this,—that we should give up the idea of its self-subsistence and conceive it as a maintained manifestation of the Divine energy."

‡ Socrates; quoted by Anderson, p. 158.

286. God, the revealer, in the objective-subjective or theological relation,—Man, the observer, the finite-subjective, the metaphysician,—and Truth, the revealed, in the subjective-objective or scientific relation, are the sole objects of philosophical inquiry. All observation should start from the observer, as the centre to which everything is referred. That which is nearest to the observer may be most closely and carefully scrutinized, and the scrutiny will furnish base lines for extending the survey to that which is more remote.

287. Man finds in himself not only the triform Intelligence, but also an analogous threefold nature,—Intelligence, Force, and a passive material frame, which is controlled by Intelligence through the instrumentality of Force. Of these three coexistences, Intelligence occupies the highest rank, and Matter the lowest, while Force is intermediate, acting and reacting between the other two.

288. This evident action and reaction has given rise to two opposite schools of philosophy: the material, which maintains that mind is a product of physical organization, and the spiritual, which, starting from the indubitable truth that we know nothing of matter except the ideas that are formed of it in our own minds, denies all material existence, and recognizes in the universe only mind and its ideas. Although the spiritual school is undoubtedly the most reasonable, a true philosophy must either recognize in the differing qualities of thought and extension sufficient grounds for admitting the substantial existence of both mind and matter, or it must satisfactorily demonstrate that all those qualities can be deduced from the admitted properties of a single form of substance.

289. The human intelligence is undoubtedly affected by material influences and associations, but that it is not a resultant of the material organization is evident from the fact that the cultivation of the material frame (beyond the mere requisites of physical health) tends rather to weaken than to increase man's power over the intellectual world,—while the cultivation of the intellect always increases his power over the material universe.

290. With whatever reason man may assign to his intelligence the highest rank in his own organization, he cannot deny that it is, in its turn, subject to a still higher external power.

291. In investigating the external world he finds, as in his own microcosm, passive matter and forces, over which his own Intelligence can exert some control, and which are more fully controlled and directed by some invisible Agency. The mode in which the Agency guides Force is called LAW. The human mind can, to a limited extent, frame laws of its own, analogous, though infinitely inferior, to those of the Supreme Agency. Human Intelligence can, moreover, investigate, comprehend, and in some cases predict the laws which govern the Universe.

292. The Supreme Agency, or Ruler of the Universe, acts, therefore, in ways which

are partially comprehensible by finite human intelligence, the comprehension being fuller in proportion as that intelligence is more highly developed,—in such ways as an Infinite Intelligence would act. And as we can conceive of no other agency than Intelligence that would be able to assign laws to force, the conclusion is irresistible that Intelligence is the highest or ruling power of the Universe. An infinite Intelligence could control an infinite Universe, but no other conceivable form of Existence would be capable of such control.

293. Many have imagined that they could conceive in some fancied necessity or fate a cause superior to Intelligence. But a rigid analysis would probably prove that the supposed necessity was only a characteristic of the Infinite Intelligence, in which the mass of mankind have always believed. For what is necessity but an attribute of rational determination? What other ground can be given for the predication of necessity than that reason declares that it must be so? What our finite reason discovers as infinitely and eternally necessary, the Infinite Reason must have declared from all eternity. Necessity, as apprehended by us, is purely ideal,—perhaps the highest of all mere ideas,—and yet the idea is doubtless the perfect transcript of the reality. Our idea could have no existence except in intelligence, and the reality is not conceivable as having any existence independent of an Infinite Intelligence.*

294. Others have supposed that Law is supreme, even Intelligence and Will being the results of organization, and therefore subject to the organic law. But what is this organic law, and what possible conception can be formed of any law, independent of a lawgiver?

295. The vital movements are almost wholly involuntary, and are generally under the control of the law of vitality, which is another name for the organic law. This law, like Will, acts for a special purpose, and its action may often be modified, and to some extent controlled, by the counteraction of the human will.

296. The same agencies (light, heat, electricity, &c.) that govern the universe in accordance with fixed laws for fixed purposes, are also used by man to accomplish his purposes in the exercise of his will.

297. The man to whom we become attached is the spiritual, not the physical man. This spiritual man controls the motions of the limbs, the organs of speech, and all the voluntary muscles, by his individual will. The will is, therefore, their law. The law, or

* If the Supreme were not intelligent and rational, it would be impossible to predict any result, or to trace out any antecedent cause. The perception of necessity is not the perception of any blind chance or fate, but the perception of a rational conclusion or law, which could only have been originally made and eternally sustained by a Rational Lawgiver.

governing power, is higher than the thing governed, and cannot, therefore, be its creature or resultant.

298. "My lawful will, simply as such, in and through itself, must have consequences, certain and without exception. . . . The idea of *Law* expresses generally nothing else but the fixed, immovable reliance of Reason on a proposition, and the impossibility of supposing the contrary.

299. "I assume such a law of a spiritual world which my own will did not enact, nor the will of any finite being, nor the will of all finite beings together; but to which my will and the will of all finite beings is subject. . . .

300. "Agreeably to what has been advanced, the law of the supersensuous world should be a *Will.*

301. "A Will which acts purely and simply as will, by its own agency, entirely without any instrument or sensuous medium of its efficacy; which is absolutely, in itself, at once action and result; which wills and it is done, which commands and it stands fast; in which, accordingly, the demand of reason, to be absolutely free and self-active, is represented. A Will, which is law in itself; which determines itself, not according to humor and caprice, nor after previous deliberation, vacillation and doubt, but which is forever and unchangeably determined, and upon which we may reckon with infallible security; as the mortal reckons securely on the laws of his world. A Will in which the lawful will of finite beings has inevitable consequences, but only their will, which is immovable to everything else, and for which everything else is as though it were not."*

302. "Now, the great complex of all this universe, in all time and all eternity, is made up of nothing more than the will of the Creator and the wills of his creatures. What is all this solid frame of sun, earth, and stars, as far as we can know anything of it, but the projection of the will of God upon the mind of man? What is all history but the action of the will of man within the limits imposed by the will of God? Will, in some form, either Divine or human, is the first principle of all existing things."†

303. "All laws, considered in the origin of their power, are despotic. *Sic volo sic jubeo.* Laws of the will are not laws which the will receives, but laws which the will gives. . . How should a law be able to produce a will?"‡

304. Necessity and Law are, therefore, subordinate to Will and Intelligence, and the analogies of Ontology, as well as the postulated Unity of Reason and the teachings of Faith, lead irresistibly to a Supreme Active Intelligence.

* Fichte.

† Solly, p. 11.

‡ Jacobi. See the third Antinomy; Kant, pp. 314–18.

CHAPTER X.

ONTOLOGICAL VIEWS.

305. BEFORE proceeding to essay a preliminary ontological analysis, it may be well to group together a few of the observations of prominent philosophers who have been led, experimentally, to conclusions analogous to those which flow necessarily from the theoretical relations of the subjective and objective.*

306. "Being and thought are therefore identical, with Parmenides. This pure thought, directed to the pure being, he declares is the only true and undeceptive knowledge, in opposition to the deceptive notions concerning the manifoldness and mutability of the phenomenal." *Schwegler*, p. 29.

307. "But my thinking, my reason is not something specially belonging to me, but something common to every rational being; something universal, and in so far as I am a rational and thinking being, is my subjectivity a universal one. But every thinking individual has the consciousness that what he holds as right, as duty, as good or evil, does not appear as such to him alone, but to every rational being, and that consequently his thinking has the character of universality, of universal validity, in a word,—of objectivity, . . . and therefore with him [Socrates] the philosophy of objective thought begins." *Id.*, p. 51, 52.

308. "Some of the ancients say that Plato was the first to unite in one whole the scattered philosophical elements of the earlier sages, and so to obtain for philosophy the three parts, logic, physics, and ethics. The more accurate statement is given by *Sextus Empiricus*, that Plato has laid the foundation for this threefold division of philosophy, but that it was expressly recognized and affirmed by his scholars, Xenocrates and Aristotle." *Id.*, p. 82.

309. "Plato distinguishes two components of the soul,—the Divine and the mortal,—the rational and the irrational. These two are united by an intermediate link which Plato calls θυμὸς, or spirit, and which, though allied to reason, is not reason itself, since it is often exhibited in children and also in brutes, and since even men are often carried away by it without reflection. This threefoldness, here exhibited psychologically, is found, in different applications, through all the last general period of Plato's literary life. Based upon the

* The selections in this chapter are taken from the works of German philosophical historians, because Kant and his successors of the modern German school have recognized a prevailing triplicity, to which they have been empirically led through the radical duality of the subjective and objective.

anthropological triplicate of reason, soul, and body, it corresponds also to the division of theoretical knowledge into science (or thinking), current opinions (or sense-perception), and ignorance; to the triple ladder of eroticism in the symposium and the mythological representation connected with this of Poros, Eros, and Penia; to the metaphysical triplicates of the ideal world, mathematical relations and the sensible world." *Id.*, p. 99.

310. Aristotle calls the soul in plants, nutritive,—in animals, sensitive; "lastly, the human soul is at the same time nutritive, sensitive, and cognitive." *Id.*, p. 129.

311. According to Aristotle, "It is by three things, therefore, nature, habit, and reason, that man becomes good." *Id.*, p. 132.

312. "To the two cosmical principles already received, viz., the world-soul and the world-reason, a third and higher one was added by the New Platonists. For if the reason apprehends the true by means of thinking, and not within itself alone; if, in order to grasp the absolute and behold the divine, it must lose its own self-consciousness, and go out beyond itself, then reason cannot be the highest principle, but there stands above it that primal essence, with which it must be united if it will behold the true. To this primal essence, Plotinus gives different names, as 'the First,' 'the One,' 'the Good,' and 'that which stands above being.' . . . In all these names, Plotinus does not profess to have satisfactorily expressed the essence of this primal One, but only to have given a representation of it. In characterizing it still further, he denies it all thinking and willing, because it needs nothing and can desire nothing; it is not energy, but above energy; life does not belong to it; neither being nor essence, nor any of the most general categories of being can be ascribed to it; in short, it is that which can neither be expressed nor thought." *Id.*, p. 155.

313. "The system of Spinoza rests upon three fundamental conceptions, from which all the rest may be derived with mathematical necessity. These conceptions are that of substance, of attribute, and of mode." *Id.*, p. 185.

314. According to Locke, "the complex ideas may be referred to three classes, viz., the ideas of mode, of substance, and of relation. . . . Our idea of substance is distinguished from all other complex ideas, in the fact that it is an idea which has its archetype distinct from ourselves, and possesses objective reality, while other complex ideas are formed by the mind at pleasure, and have no reality corresponding to them external to the mind. We do not know what is the archetype of substance, and of substance itself we are acquainted only with its attributes. A relation arises when the understanding has connected two things with each other in such a way, that in considering them, it passes over from the one to the other." *Id.*, p. 196.

315. "God gives us ideas; but as it would be contradictory to assert that a being could give us what it does not possess, so ideas exist *in God*, and we derive them from

Him. Those ideas in God may be called archetypes, and those in us ectypes. In consequence of this view, says Berkeley, we do not deny an independent reality of things; we only deny that they can exist elsewhere than in an understanding." *Id.*, pp. 221–2.

316. "Resting on the perception that there are within the soul two faculties, one of knowing and one of willing, Wolff divides philosophy into two great parts,—theoretical philosophy (an expression, however, which first appears among his followers), or metaphysics, and practical philosophy. Logic precedes both, as a preliminary training for philosophical study. Metaphysics are still farther divided by Wolff into ontology, cosmology, psychology, and natural theology; practical philosophy he divides into ethics, whose object is man as man; economics, whose object is man as a member of the family; and politics, whose object is man as a citizen of the state." *Id.*, p. 224.

317. [Kant.] "All the faculties of the soul, he says, may be referred to three, which are incapable of any farther reduction; knowing, feeling, and desire. The first faculty contains the principles, the governing laws for all the three. So far as the faculty of knowledge contains the principles of knowledge itself, is it theoretical reason, and so far as it contains the principles of desire and action, is it practical reason, while, so far as it contains the principles which regulate the feelings of pleasure and pain, is it a faculty of judgment. Thus the Kantian philosophy (on its critical side) divides itself into three criticks; (1) Critick of pure, *i. e.* theoretical reason; (2) Critick of practical reason; (3) Critick of the judgment." *Id.*, pp. 237–8.

318. "The faculty of judgment is the middle link between the understanding as the faculty of conceptions, and the reason as the faculty of principles. . . . The object of the faculty of judgment is, therefore, the conception of *design* in nature; for the evidence of this points to that supersensible unity which contains the ground for the actuality of an object. And since all design and every actualization of an end is connected with pleasure, we may farther explain the faculty of judgment by saying, that it contains the laws for the feeling of pleasure and pain." *Id.*, p. 262.

319. "The positive philosophic views which Jacobi exhibits in this treatise ['On the Doctrine of Spinoza, in letters to Moses Mendelssohn'], can be reduced to the following three principles: (1) Spinozism is fatalism and atheism. (2) Every path of philosophic demonstration leads to fatalism and atheism. (3) In order that we may not fall into these, we must set a limit to demonstrating, and recognize faith as the element of all metaphysic knowledge." *Id.*, p. 272.

320. "A theory of science must posit some supreme principle, from which every other must be derived. This supreme principle must be absolutely, and through itself, certain. . . . . Its test and demonstration can only thus be gained, viz., if we find a principle to which all science may be referred, then is this shown to be a fundamental principle. But

besides the first fundamental principle, there are yet two others to be considered, the first of which is unconditioned as to its content, but as to its form, conditioned through and derived from the first fundamental principle; the other the reverse. The relation of these three principles to each other is, in fine, this, viz., that the second stands opposed to the first, while the third is the product of the two. Hence, according to this plan, the first absolute principle starts from the Ego, the second opposes to the Ego a thing, or a Non-Ego, and the third brings forward the Ego again in reaction against the thing, or the Non-Ego. This method of Fichte (thesis,—antithesis,—synthesis), is the same as Hegel subsequently adopted and applied to the whole system of philosophy, a union of the synthetical and analytical methods." *Id.*, p. 285.

321. "Schelling thus distinguishes the two sides of philosophy. All knowledge rests upon the harmony of a subject with an object. That which is simply objective is natural, and that which is simply subjective is the Ego or intelligence. There are two possible ways of uniting these two sides: we may either make nature first, and inquire how it is that intelligence is associated with it (natural philosophy), or we may make the subject first, and inquire how do objects proceed from the subject (transcendental philosophy). The end of all philosophy must be to make either an intelligence out of nature, or a nature out of intelligence. . . Both, however, are only the two poles of one and the same knowledge, which reciprocally attract each other; hence, if we start from either pole, we are necessarily drawn towards the other." *Id.*, p. 318.

322. [Hegel.] "Mind is at first theoretical mind, or intelligence, and then practical mind, or will. It is theoretical in that it has to do with the rational as something given, and now posits it as its own; it is practical in that it immediately wills the subjective content (truth), which it has as its own, to be freed from its one-sided subjective form, and transformed into an objective. The practical mind is, so far, the truth of the theoretical. The theoretical mind, in its way to the practical, passes through the stages of intuition, representation, and thought, and the will on its side forms itself into a free will through impulse, desire, and inclination. . . .

323. "This stand-point [of moral reflection] has three elements: (1) the element of resolution (*vorsatz*), where we consider the inner determination of the acting subject, that which allows an act to be ascribed only to me, and the blame of it to rest only on my will (imputation); (2) the element of purpose, where the completed act is regarded not according to its consequences, but according to its relative worth in reference to myself. The resolution was still internal; but now the act is completed, and I must suffer myself to judge according to the constituents of the act, because I must have known the circumstances under which I acted; (3) the element of the good, where the act is judged according to its universal worth. The good is peculiarly the reconciliation of the particular

subjective will with the universal will, or with the conception of the will; in other words, to will the rational is good." *Id.*, pp. 358, 360.

324. "We regard, indeed, generally the three ideas, God, freedom, and immortality, as the chief subject-matter, or content of philosophy." *Chalybäus*, p. 4.

325. [Kant.] "These three ideas (soul, world, deity), would thus furnish the principles of the three divisions of metaphysics, namely, rational psychology, cosmology, and theology. . . .

"Since then, it has been made out, as a conseqence of the Critick of Pure Reason, that the proper objects of metaphysics, namely, God, universe, and mind (freedom, subjective being), are wholly inaccessible to our cognition, and lie beyond the limit of all philosophical *knowledge*, . . we cannot indulge the least hope of ever learning, by the help of speculation, whether or not there are transcendent beings that correspond to these ideas." *Id.*, pp. 34, 48.

326. "While with Hegel, in his logic, nature-philosophy and philosophy of mind, we everywhere encounter a tripartite system, bound together by a single, formal, and real principle, and conditionated by one decisive method, which repeats itself in a rhythmic and symmetrical manner throughout; so also in Herbart's system we find, it is true, such a threefold division,—but one which, apart from many other considerations, differs wholly from that of Hegel's system in this, that the three cardinal divisions, or the Logic, Metaphysics, and Æsthetics, are neither bound together by a common, real, or formal principle, nor do they acknowledge, as presiding over them, any general and fundamental doctrine, which might contain and determine such a fundamental principle." *Id.*, p. 83.

327. "According to Herbart, we can think of the change as taking place in a threefold manner, either as proceeding from external causes, or by self-determination, or finally as absolute origination or becoming." *Id.*, p. 100.

328. "So far as philosophy busies itself with the forms of thought as such, it is logic; so far as it penetrates in thought the content that is given us, thus cognizes the being and elevates it to knowledge, it is metaphysics, which is the fundamental science of philosophy. Schleiermacher, however, includes the two parts of it, the metaphysical division of the general doctrine of cognition, and the special logical, together under the name of *Dialectick*. Under this general division comes in due order everything that is conceivable, consisting upon the one hand, of nature, upon the other, of the sphere of conscious action, consequently of physics and ethics, so that on the whole, the ancient division of philosophy into dialectick, physics, and ethics is re-established." *Id.*, pp. 191–2.

329. [Schelling.] "In this freedom [of the ideal] it was said that we encounter the last potentializing act, whereby the whole of nature became transfigured into sensation, intelligence, and finally into will. In the last and highest instance, there is no other

being whatever than volition. Volition is primordial being, and with this alone all its predicates of groundlessness, independence of time, and self-affirmation conform." *Id.*, p. 265.

330. [Hegel.] "Universality, speciality, and individuality, are accordingly the three momenta of the idea, and are present in it as an unity. . .

"We are consequently suddenly withdrawn from the sphere of the subjective logic, and transported into the region of objectivity, or into the '*doctrine of the object*,' which resolves itself into 'mechanism, chemism, and teleology.'" *Id.*, pp. 335, 338.

331. "The Hegelian fundamental schema,—Being, Naught,* Origination,—does not correspond to the schema of objective teleology,—principle, means, and effect; but the origination or process, *i. e.* the means, is interposed as an eternal self-mediation in the place of the purpose." *Id.*, p. 381.

332. "We have in the present work traversed but a comparatively small, although rich, division of the whole development of Philosophy,—in short, its last or modern phase only; and have seen in this that the chief business of human thought is and must be to discover principle, means, and end, both in the singular and in the whole. All three moments ought to be one or united; but they must also be distinguished, and each in its own place must necessarily be that to which, by this place, it is entitled or justified." *Id.*, p. 385.

333. In all the foregoing quotations, it is easy and interesting to trace the influence of the great idea of relativity, and in nearly every instance the idea is plainly developed under the forms of direction to, in, or from some assumed subjective or objective centre, although, in consequence of the experimental nature of the development, the boundaries of the several forms are not as clearly marked as they would have been, if the theoretical limitation had been thoroughly understood, and constantly kept in view.

CHAPTER XI.

DEDUCTION OF THE KANTIAN CATEGORIES.

334. In attempting to reach the *summa genera* of the knowable, we may start either from the objects of thought, or from thought itself. The results of the two processes will

* All our reasoning about the Absolute must be, as Mansel well observes (p. 85), not about "the nature of the Absolute in itself, but only our own conception of that nature. The distortions of the image reflected may arise only from the inequalities of the mirror reflecting it." If we imagine the Absolute to be not only independent of relation, but absolutely devoid of relation, our conception of it, like Hegel's, must be simply Naught.

naturally differ, but the difference should not be irreconcilable, and philosophy can never make much progress until a reconciliation is effected. "Aristotle attempted a synthesis of things in their multiplicity,—a classification of objects real, but in relation to thought; —Kant, an analysis of mind in its unity,—a dissection of thought, pure, but in relation to its objects. The predicaments of Aristotle are thus objective, of things as understood; those of Kant subjective, of the mind as understanding. The former are results *a posteriori*,—the creations of abstraction and generalization; the latter, anticipations *a priori*,—the conditions of those acts themselves."*

335. The method of Aristotle was nearly perfected by its author, and for more than two thousand years, his disciples have endeavored in vain to extend or improve his system. The method of Kant was merely initial, and the revival of philosophy during the past century has shown that it was productive. It has the advantage of starting from that which is best known, the subjective, in its endeavors to learn the unknown, while Aristotle started from the objective, of which he was obliged to assume a reality for which he had only subjective evidence.

336. Without attempting to harmonize the objective and subjective categories,—without even endeavoring to give the best philosophical explanation of the basis on which either of them rests,—it may reasonably be expected that a subjective symbolism should at least show a possible mode, if not the best mode, of accounting for Kant's empirical arrangement of the subjective categories. Let us see with what success we can deduce them from the relations of the primitive forms of Consciousness.

337. All analysis proceeds from the general to the particular that is embraced under it. Our highest general idea of the mind, is the idea of Consciousness, and the first question that suggests itself for our analysis is, How does Consciousness regard the objects of its cognition, or in what different modes can it consider them? The answer will naturally be sought in accordance with the conditions of Motivity, Spontaneity, and Rationality.

338. Motivity, although it refers to objects exterior to ourselves, cannot immediately give us those objects. It relates only to phenomena, and to the influence of those phenomena on our own minds. If, for example, I receive a sensation of solidity, or heat, or color, the *sensation* is entirely subjective; it belongs exclusively to myself, and not to the body, to which Motivity refers as its cause. I cannot, therefore, merely as receptive, assert the reality of anything objective; the most I can do is to admit its Possibility.

339. Spontaneity, being exclusively subjective in its action as well as its reference, is entirely valid in all its determinations. I know absolutely all that I feel, wish, do, or think, and hence I derive a consciousness superior to the mere possibility of the Motivity,—

* Hamilton: *Discussions*, p. 32.

a consciousness of Reality. My own reality is more evident than that of any being out of myself, and the highest reality to which I can attain is, therefore, that of a Spontaneous Intelligence.

340. Rationality decides not only with unvarying uniformity from the data that are given it, but it does so with the full conviction that it would be impossible for any intelligent being to decide otherwise from the same data. In viewing the possibility of Motivity, it decides that there must necessarily be an external objective cause of all our external impressions; it seeks in the reality of the subjective Spontaneity a necessary object for its consideration; and it conjoins the idea of necessity with all its determinations, thus completing the circle of our modes of thought.

341. Consciousness, therefore, in the three conditions of intelligence, gives us the three categories of Modality,—Possibility, Reality, and Necessity, all of which refer to General Science.

342. Subjecting the several mental states in turn to the same kind of analysis, our next inquiries must be: How do Motivity, Spontaneity, Rationality regard the objects of their cognition? We will seek the answers by the same clue that we adopted in the case of Consciousness.

343. A passive and comparatively quiet state of mind is the earliest, the easiest, and perhaps the most common at all periods of life. The facts of Motivity are therefore the most evident, the most generally admitted, and the most readily understood. Motivity is emphatically the faculty of childhood or pupilage; and all its teachings are received with the implicit faith of the child and pupil.

344. All our impressions are susceptible of increase or diminution. Hence through Motivity we readily obtain the idea of Quantity.

345. All the impressions of mere Motivity are single and momentary. Merely as receptive beings, we neither distinguish parts of objects nor unite different impressions together. If we feel, our sensation is a unit,—merely a feeling, and nothing more; if we see, we see an object as a unit, and so with every impression on the senses. It is merely the impression that is cognized through Motivity, and the category of Motivity, cognizing its own impressions, is, therefore, Unity.

346. Spontaneity unites several determinations in its own consciousness. It embraces the faculty of attention, and applying itself to the determinations of Motivity, it can attend successively to all the parts of an object or of an impression, and derive the idea of plurality from unity. The category of Spontaneity, cognizing the impressions of Motivity, is, therefore, Plurality.

347. The office of Rationality is, as we have seen, to cognize and compare the representations of the other intellectual conditions. Applying itself to the determinations of

Motivity, it will therefore recognize both the receptive unity and the spontaneous plurality, and from their relation will derive the category of Totality, in which Unity and Plurality are both combined.

348. Motivity, therefore, in the three conditions of intelligence, gives us the three categories of Quantity,—Unity, Plurality, and Totality, all of which refer, through the motive category of Modality, to the Science of the Possible, and particularly to Mathematical Science.

349. Next in order of prominence as well as of acquisition, are the determinations of Spontaneity. The subject cannot regard itself otherwise than as object, and there is therefore more difficulty attending the study of the purely subjective, than we have found in the objective, as manifested through Motivity.

350. Spontaneity is the active, laboring state of the mind, corresponding to the vigor of youth and early manhood. In investigating its laws, one of the first inquiries is, What influence do we have by our voluntary action over the objects of our cognition, or what quality do we communicate to them?

351. All impressions of Spontaneity on Motivity, are real or affirmative. We have, indeed, no power of limitation or of negation, as mere receptive beings, but we assert fully and positively, every impression that we receive. Thus, we can never admit that the senses deceive us, for the senses are merely media for conveying impressions. We exercise our attention or spontaneity, together with our senses or motive-rationality, and an impression follows, the reality of which is undoubted. A further effort of spontaneity is required to interpret the meaning of that impression, and if that secondary effort is insufficient, we are led into error. The category of Motivity, cognizing the impressions of Spontaneity, is therefore Affirmation.

352. The peculiar office of Spontaneity, particularly when concerned with its own actions, or with Reality, of which its actions are the representatives, is to define, limit, and give precision to our ideas.

353. We have seen that what are called delusions of the senses, are properly errors of spontaneity. I have, for example, the impression of an object on the optic nerve. Of the reality of the impression there can be no doubt, but great care may be necessary to give it its proper interpretation. I must first inquire whether the impression is occasioned by a disease of the nerve, or by the normal stimulus of an external object. If I am satisfied that it proceeds from an outward object, I must then attend to the angles of vision which determine the outline, the modulations of light and shade that indicate the form, the distinctness or indistinctness that mark its relative nearness or distance, and enable me to judge of its size, the clearness or haziness of the atmosphere, sharpness or obscurity of vision, and any other circumstances that may affect my decision. A failure of proper at-

tention in any of these limiting particulars, may be the source of error. The category of Spontaneity, cognizing its own action, is therefore Limitation.

354. Rationality, comparing the receptive affirmation and the spontaneous limitation, can alone set impassable limits, and give us the category of Negation. The child, in beginning to gratify his desire for knowledge, accepts everything that is told him with implicit faith. It is only after Rationality has acquired considerable development, that he begins to doubt or deny.

355. Spontaneity, therefore, in the three conditions of intelligence, gives us the three categories of Quality,—Affirmation, Limitation, and Negation; all of which refer, through the spontaneous category of Modality, to the science of Reality, or Existence, and particularly to Natural Science.

356. Rationality is not only the highest of our faculties, but it is the latest developed, and the most rarely found in full development. It holds the same rank in the human mind that man himself occupies in the created world, and as reasoning beings are superior to beings of impulse and instinct, so is the reasoning man, the man who seeks and loves the true and perfect, superior to the man of appetite and passion.

357. A thorough investigation of Rationality is therefore, as we might naturally suppose, and as we shall find in the pursuit of our inquiries, attended with greater difficulties than the study of Motivity or Spontaneity. Some of these difficulties are indeed insurmountable by finite and progressive beings, for from its very nature, reason requires the perfect and infinite for its full satisfaction. So long, therefore, as there is any leaven of imperfection in us, we can only approximate nearer and nearer to its ends, without ever attaining them.

358. The first step in our progress, the investigation of the rational categories, is, however, comparatively easy, guided as we may be by the results we have already obtained. We need only inquire (since the objective becomes ideally represented to us only through Relation), What are Motivity, Spontaneity, and Rationality, and what relations do they severally indicate?

359. Motivity, as we have seen, refers merely to phenomena, from which we obtain ideas of the accidental qualities of bodies. Rationality asserts the necessity of some reality in which those qualities inhere. The category of Motivity in Rationality is, therefore, that of Inherence and Subsistence, or Substance and Accident.

360. Spontaneity is the faculty of action. Rationality assigns the negative limit to action, which is reaction. The relation of the two gives us the category of Spontaneity in Rationality, which is that of Action and Reaction.

361. Rationality, as will be more evident hereafter, is the faculty of Cause. The correlative of cause is effect. The same relation is easily discovered by comparing the two

preceding categories. The category of Rationality in Rationality is, therefore, that of Cause and Effect.

362. Rationality, therefore, in the three conditions of intelligence, gives us the three categories of Relation: Substance and Accident, Action and Reaction, and Cause and Effect, all of which refer, through the rational category of Modality, to the science of the Necessary, and particularly to Metaphysical Science.

363. The twelve categories thus deduced, correspond with the categories of Kant, except in the arrangement of the subdivisions of Quality and Relation. Kant could discover no reason for his arrangement, or for the precise number of categories that he propounded,* but his mind was eminently analytical, and proceeding in strict accordance with the necessary laws of analysis, his researches were crowned with a success nearly as complete, as if he had fully perceived the dependence of the result upon those laws. Perhaps in no portion of his great works is his genius more evident, than in his development of the laws of perceptive unity, with no other guides than his own discernment, and the meagre clue afforded by the categories of Aristotle.†

364. So far as the categories are the representatives of ideas that we have received, they belong to our Motivity, and therefore have an objective reference as to their origin. If we apply them to the subject through Spontaneity, or to general judgments through Rationality, they undergo a formal modification that can be readily discerned.

365. If we represent the judgment forms by means of symbols, their mutual relation and deduction will be more evident, and the defects of our nomenclature, whatever they may be, will disappear in the symbolic formula. In the following table, X is the symbol of the categorical or rational forms; Y of the subjective or spontaneous; Z of the condi-

* "But respecting the property of our understanding, to effect unity of apperception *à priori*, only by means of the categories, and precisely only in this manner and the number thereof, no more motive can be adduced than why we have exactly these, and no other functions of judgment, or why time and space are the only forms of our possible intuition." Kant, pp. 96–7.

† Aristotle's categories were, Being, *οὐσίαν*; Quantity, *ποσὸν*; Quality, *ποιὸν*; Comparison or Relation, *πρός τι*; Where, *πόθι*; When, *ποτὲ*; Posture, *κεῖσθαι*; Having, *ἔχειν*; Action, *ποιεῖν*; Passion, *πάσχειν*. Vol. I, p. 20. Hamilton arranges these categories as follows: "*Being by itself* corresponds to the first category of Aristotle, equivalent to substance;—*Being by accident* is viewed either as absolute or as relative. As absolute, it flows either from the matter, or from the form of things. If from the matter, it is *Quantity*, Aristotle's second category; if from the form, it is *Quality*, Aristotle's third category. As relative, it corresponds to Aristotle's fourth category, *Relation*; and to Relation all the other six may be reduced." *Logic*, p. 141.

There are six simple ideas, "according to the most accredited opinion, in the school of the Nyaya [founded by Gotama]. These are substance, quality, action, the common (the general, genus), property (species, the individual), and relation. Some authors add a seventh element,—privation or negation; others add two more still,—power and resemblance." *Cousin: Hist. of Mod. Phil.*, Vol. I, p. 382.

tional or motive. The modification of the several forms under their relation to the general Consciousness, is denoted by C; M symbolizes the relation to Motivity; S, to Spontaneity; R, to Rationality.

366. With the interpretation of the categorical forms, we are already familiar. If we were to interpret the others in a similar manner, the entire table would be nearly as follows:

X, Categorical.	*Y, Subjective.*	*Z, Conditional.*
XM, Quantity.	YM, Quantity.	ZM, Quantity.
XMM, Unity.	YMM, Complexity.	ZMM, Universal.
XMS, Plurality.	YMS, Simplicity.	ZMS, Particular.
XMR, Totality.	YMR, Aggregation.	ZMR, Individual.
XS, Quality.	YS, Quality.	ZS, Quality.
XSM, Affirmation.	YSM, Admission.	ZSM, Affirmative.
XSS, Limitation.	YSS, Qualification.	ZSS, Qualitative.
XSR, Negation.	YSR, Enclosure.	ZSR, Infinite.
XR, Relation.	YR, Relation.	ZR, Relation.
XRM, Substance and Accident.	YRM, Intelligence and Manifestation.	ZRM, Subject and Predicate.
XRS, Action and Reaction.	YRS, Restraint and Resistance.	ZRS, Condition and Conditioned.
XRR, Cause and Effect.	YRR, Design and End.	ZRR, Foundation and Consequence.
XC, Modality.	YC, Modality.	ZC, Modality.
XCM, Possibility or Impossibility.	YCM, Hypothesis.	ZCM, Problematical.
XCS, Reality or Non-Entity.	YCS, Faith.	ZCS, Assertive.
XCR, Necessity or Contingence.	YCR, Knowledge.	ZCR, Axiomatic.

367. The number of possible categories, like the number of possible faculties, is infinite. The extent to which the subdivision should be carried, must depend entirely on the purposes we wish to serve.

CHAPTER XII.

APPLICATION OF CATEGORIES,—SPACE, TIME, AND POSITION.

368. Having now indicated the groundwork of science, and having shown that upon this groundwork a superstructure may be erected of infinite detail, we shall confine ourselves principally in the farther execution of our plan, to a partial development of the categories of Relation and Modality, as applied to objects of cognition.

369. Knowledge may be either modally absolute, real, or problematical.* Of absolute knowledge, we have an example in pure mathematics, and in every axiom, or proposition which carries with itself the perception of its necessity and universal validity. Real knowledge embraces every fact which we are compelled to believe by the constitution of our minds, but of which we do not perceive the entire necessity. Problematical knowledge, or belief, covers everything which we believe to be true, but the truth of which depends on circumstances which it is impossible for us to determine with certainty.

370. Science, properly so called, is concerned principally with the absolute or necessary, and the real. General science is based upon, and includes all the necessity that is discerned by the intelligence in its several conditions.

371. Our cognitions in their reference, as we have seen, are either objective, subjective, or ideal. The objects of our cognition, or the things cognized, may be viewed in three states, analogous to the three conditions of consciousness. We may regard them either as passive, active, or sustaining.†

372. There can be no possibility, except in accordance with reality and necessity. If objects can have a passive, an active, or a sustaining existence, there must be some reality that renders their existence possible. In attempting to ascertain the forms of that reality, by analyzing the objective, we enter on a task both delicate and fruitless, unless that Greatest, Wisest, and Best, toward which all philosophy aspires as the necessary goal of its inquiries, is pure Intelligence, and therefore purely subjective, and what is objective to our finite intelligence, is in reality only a form or product of a higher subjective.‡ We

* The term knowledge has been confined by some writers to absolute truth. But we speak of the acquisition of knowledge, including what we learn from books and from testimony, and the more extended definition here recognized therefore corresponds with the common acceptation of the term.

† Every cognition must be either of the *not-me*, or objective,—of the *me*, or subjective,—or of the ideal, which embraces every logical antecedent of either objective or subjective manifestations. The mind, as motive, is nearly passive; as spontaneous, active; as rational, sustaining or authoritative,—every effort of rationality being an effort to approximate to the underlying, upholding, substantial, or necessary.

‡ "Every cogitative faculty, though it is not the sole cause of its own immediate (apparent) object, yet has a share in making it: thus the eye or visive faculty hath a share in making the colors which it is said to see; the ear or auditive power, a share in producing sounds, which yet it is said to hear; the imagination has a part in making the images stored in it; and there is the same reason for the understanding, that it should have a like share in forming the primitive notions under which it takes in and receives objects; in sum, the immediate objects of cogitation as it is exercised by men, are *entia cogitationis*, all phenomena; appearances that do no more exist without our faculties in the things themselves, than the images that are seen in water or behind a glass, do really exist in those places where they seem to be." *Essay upon Reason and the Nature of Spirits*, by Richard Burthogge, M.D., quoted by Solly, p. 285.

If we could fully comprehend the share of the Will in making its own motives, we could doubtless better understand the extent and limits of free will.

have already discovered good reasons for believing that such is the case, and we shall find, if we investigate thoroughly, that on any other hypothesis, all pretended philosophy,—as well as all supposed science, whether it be self-styled positive, or speculative,—is an idle dream.

373. As the objective mode of Existence embraces everything as it is in itself, independent of any (finite) subjective relations or modifications, it corresponds to the *essentia* of Cicero, and it may, therefore, perhaps be best designated by the term ESSENCE.

374. The Necessary modification of Essence, which renders all other manifestations possible, we will call FORM, inasmuch as it determines the qualities or characteristics that must attend every conceivable form of essential being.

375. To Reality in Essence, we may apply the name of SUBSTANCE, in accordance with the general usage of philosophers, though some, like Xenophanes, Descartes, and Spinoza, so limit the meaning of substance, that the term is nearly synonymous with Absoluteness, or Self-Existence, and is therefore only applicable to the Deity.

376. The Possible in Essence, as the source of all the modifications of which it is susceptible, we will call CONDITION.

377. In extending this analysis, in order to determine in the first place the subdivisions of Form, the analogy that we have followed hitherto, justifies the use of such assistance as we can derive from the primary forms of Consciousness.

378. Through the mediation of Motivity, we obtain ideas of difference, externality, extension, and of Space, in which extension is alone possible, and in which is also included the possibility of passive existence.

379. Through Spontaneity, we obtain ideas of intelligent action, succession, duration,—and of Time, which renders them all possible.

380. Through Rationality, as the faculty of precise determination, we obtain ideas of relation, cause, fundamental being,—and of that which renders them all possible, which we will call Position.*

381. Position bears the same relation to Space and Time, as the Rational holds in all cases to the Motive and the Spontaneous. It embraces not merely place and date, but also limit, relation, diversity, multiplicity, law, and all the determinations which fix the boundaries of any conceivable form of being,† in Space or Time.

* There is a curious etymological connection, that deserves a passing notice, between *lay*, *law*, *p-lac-e*, *λεγ-*, *λογ-*, *lec-*, *loc-*.

† "I must tell you at once that human reason, in whatever manner it is developed, however occupied, whether it stop at the observation of this nature which surrounds us, or whether it dart into the depths of the interior world, conceives all things only under the condition of two ideas. Does it examine numbers and quantity? it then sees nothing but unity or multiplicity. . . . Does it occupy itself with space? it can consider it only under two

382. Space and Time are in themselves infinite, and forms or manifestations or attributes of the Absolute Infinite. In them the indefinite and the infinite are as nearly equivalent as it is possible to suppose them, and our indefinite conceptions are so nearly adequate, that we need hardly desire them to be more complete. But without Position, both Space and Time would be empty voids,—not only void of reality, but also void of conceivability.

383. When Hamilton says that Time and Space are only the images or intuitions or concepts "of a certain correlation of existences,—of existence, therefore, *pro tanto, as conditioned*,"* he appears to have his mind fixed on Position, rather than on the Infinites that make Position possible. Time is the absolute (of Cousin) which renders possible "the image or concept of a certain correlation of existences," which is *date*, and not time. Space is the parallel absolute which renders possible *place*, or conditioned Space.

384. In Position, the finite and infinite are harmoniously blended. Each single or conditioned position is finite, but position regarded as unconditioned, or in its entire possibility, is as infinite as Space and Time. By means of position, we obtain the ideas of relative infinites which constitute our indefinite concepts of absolute infinites. In all infinites, it is possible, through position, to distinguish semi-infinites,—infinite at one extremity, and finite at the other.

385. It should always be borne in mind, that in every cognition, the three forms of Consciousness co-operate, and it is impossible for us to assign the precise limit of each. How far Space, Time, and Position, are objective, subjective, or rational, it is not necessary for us to determine; if we perceive their reality under each phase of cognition, it is sufficient for the purposes of our analysis.†

points of view: it conceives a space determinate and limited, or the space of all particular spaces, absolute space. . . . Does it consider things under this single relation, that they exist? it can conceive only the idea of absolute existence, or of relative existence. . . . In the moral world, does it perceive anything beautiful or good? it then irresistibly transports this same category of the finite and the infinite, which becomes the imperfect and the perfect, the ideal beauty, and the real beauty, virtue with the miseries of reality, or holiness in its exaltation, and in its unsullied purity." *Cousin: Hist. of Mod. Phil.*, Vol. I, p. 76–7.

* *Discussions*, pp. 35, 36.

† I was led to infer the necessary reality of a third form, from the three necessary relations of the subjective. After I had decided upon the name Position, I found that Mansel had also fixed upon three "laws or formal conditions of experience." He says (p. 184), "Of these conditions, I have in a former lecture enumerated three,—Time, Space, and Personality; the first as the condition of human consciousness in general; the second and third as the conditions of the same consciousness in relation to the phenomena of matter and of mind respectively." Personality appears to be a modification of Position, that supposes some mental manifestation; but in entire independence of all manifestation, Time, Space, and Position have a necessary existence. Mahan (p. 218), enumerates as logical antecedents of phenomena, "the ideas of time, space, substance, personal identity, and cause." The last three of these ideas may all be ranked under Position.

386. Space and Time have been considered by some writers as mere forms of thought, having no existence in themselves independent of our own minds. So far as they relate to our cognitions and render them possible, this view is correct; but so far as they relate to the objects of our cognitions, and render them possible, their objective reality must be as entire, as that of the objects which they embrace.* If they were indeed mere subjective forms, their existence would of itself be sufficient evidence of the Eternal and Infinite existence of an Intelligent Being, and in the difficulty that many of the most profound investigators find, in assigning to them any other than a subjective existence, we have another evidence of the fundamentally subjective nature of all Being.

387. Kant, in consequence of the subjective character of his whole system of philosophy, attended mainly to the subjective phase of Space and Time, and because their ideas, as they exist in the mind, are adequate,—or in other words, because objective space and time have precisely those properties which are fully embraced in our subjective ideas, and no others,—he may, perhaps, have sometimes been led to doubt their objective reality. His language, at least, is such as to afford a plausible justification to those of his successors who have denied such reality, and thus involved themselves in endless confusion and mystification. As Kant is often quoted, in defence of reasoning which is capable of such perversion, it may be well to quote somewhat largely from his remarks on Space, in order to ascertain, as nearly as possible, what his views really were.

388. "By means of the external sense (a property of our mind), we represent to ourselves objects as external to us, and these all in space. . . The internal sense, by means of which the mind envisages itself or its internal state, gives indeed no intuition of the soul itself as an object; but there is still a determinate form, under which the intuition of its internal state alone is possible, so that all which belongs to the internal determinations is represented in relationships of Time. Externally, Time can be viewed as little as Space, as something in us. Now what are Time and Space? Are they real beings? Are they in fact only determinations, or likewise relations of things, but still such as would belong to these things in themselves, though they should not be envisaged; or are they such, that they cleave only to the form of the intuition, and consequently to the subjective property of our mind, without which these predicates could not be attributed even to anything. . . .

389. "1st. Space is no empirical conception which has been derived from external ex-

* Even Hamilton says (*Discussions*, p. 572), "It is one merit of the philosophy of the Conditioned, that it proves space to be only a law of thought, and not a law of things." Is this true? Is not space a law of things material? Although the soul acts in space, we do not necessarily think of it as occupying any definite place. If space is a law of thought, it is so far a law of mind, and it would seem to be still more a law of such forms of reality as can exist only in space.

periences. For in order that certain sensations may be referred to something external to me (that is, to something in another part of space to that in which I am), and likewise in order that I may be able to represent them as without of and near to each other, consequently not merely different, but as in different places, the representation of space* for this purpose must already lie at the foundation. The representation of space cannot therefore be borrowed from the relations of the external phenomenon by experience, but this external experience is itself first only possible by the stated representation.

390. "2d. Space is a necessary representation *à priori*, which lies at the foundation of all external intuitions. We can never make to ourselves a representation of this,—that there is no space,—although we may very readily think that no objects therein are to be met with. It is therefore regarded as the condition of the possibility of phenomena, and not as a determination depending upon them, and it is a representation *à priori*, which necessarily lies at the foundation of all external phenomena.†

391. "3d. Space is no discursive, or as we may say, universal conception of the relationships of things in general, but a pure intuition. For in the first place, one can only figure to oneself, one space, and when we speak of several spaces, we then understand by this only parts of one and the same single space. These parts too, could not precede the sole all-embracing space, as if constituent parts of the same (whence its aggregate is possible), but only in it can they be thought. It is essentially one,—the diversity in it, consequently also the universal conception of spaces in general rests solely upon limitations. Hence it follows, that in respect of it, an intuition *à priori* (which is not empirical), lies at the foundation of all conceptions of it. And thus all geometrical propositions, for example this: 'That in a triangle, two sides together are greater than the third,' never could be deduced from the general conceptions of line and triangle,‡ but from intuition, and certainly *à priori*, with apodictical certainty.

392. "4th. Space is represented as an infinite given quantity. We must, indeed, think each conception as a representation which is contained in an endless multitude of different possible representations (as their common sign); consequently it contains these in itself; but no conception as such can be so thought, as if it contained an infinite§ multitude of

* Observe that Kant does not say space itself, but merely "the representation of space."

† I am unable to reconcile this second clause with any belief which does not recognize an objective reality of space. Kant speaks of the *representation* of space as something subjective, and a representation seems necessarily to imply a thing represented. The admission that we can make no representation of the non-being of space, places its existence among the fundamental self-evident faiths that cannot be rejected without annihilating all certainty. Every one who admits both an objective and a subjective phase of phenomena, must also admit an objective as well as a subjective Space, "as the condition of the possibility of phenomena."

‡ The author's meaning is here somewhat obscure and questionable.

§ The ambiguity of the word *infinite*, invalidates this whole argument.

representations in itself. Nevertheless, space is so thought (for all parts of space are infinitely coexistent); consequently, the original representation of space is *Intuition à priori*, and not *Conception*. . . . .

393. "Now, how can an external intuition dwell in the mind, which precedes the objects themselves, and in which intuition the conception of these last may be determined, *à priori?* Evidently not otherwise than so far as it (*intuition*) has its seat merely in the subject, as the formal property of this (*subject*) being affected by objects, and thereby of receiving *immediate representation* of them; that is, *Intuition*, consequently, only as form of the external *sense* in general. . . .

"*Conclusions from the above Conceptions.*

394. "1st. Space represents no property at all of any things in themselves, nor does it represent them in their relationship to each other; that is, it represents no determination of them which attaches to the objects themselves, and which remains if we also make abstraction of all the subjective conditions of intuition. For neither absolute nor relative determinations can be envisaged before the existence of the things to which they belong, nor consequently *à priori.**

395. "2d. Space is nothing else but the form only of all phenomena of the external senses,—that is, the subjective condition of sensibility, under which alone external intuition is possible to us. Now, since the receptivity of the subject to be affected by objects necessarily precedes all intuitions of these objects, it may be understood how the form of all phenomena can be given in the mind previous to all real perceptions, consequently *à priori;* and how this, as a pure intuition, in which all objects must be determined, can contain principles of their relationships prior to all experience.

396. "We can, therefore, only from the point of view as men, speak of Space, Extended Beings, &c. If we depart from the subjective condition under which we alone can receive external intuition, that is to say, the way we may be affected by objects, the representation of space then means nothing. This predicate is only so far applied to things as they appear to us,—that is, as they are objects of sensibility. The constant form of this receptivity, which we name sensibility, is a necessary condition of all relationships wherein objects are envisaged as external to us, and if we make abstraction of these objects, it is a pure intuition which bears the name of Space. As we cannot make the particular conditions of sensibility into the conditions of the possibility of things, but only of their phenomena, we may very well say that space comprehends all things that may appear to us externally, but not all things in themselves,—whether they can or

* This is a *petitio principii,* as Kant himself admits with regard to *subjective* space.

cannot be envisaged,—or by whatever subject we choose. . . . If I join in this case the condition of the conception, and say 'all things as external phenomena are coexistent in space,' this rule is valid universally and without restriction. Our exposition, consequently, teaches the *Reality* (that is the objective validity) of space in reference to all that externally as object can be presented to us, but at the same time the *Ideality* of space, in reference to things, if they are considered in themselves by means of reason,—that is, without regard to the nature of our sensibility. We maintain, therefore, the *empirical reality* of Space (in respect to all possible external experience), although, indeed, we acknowledge the *transcendental ideality* of the same,—that is, that it is nothing,—so soon as we omit the condition of the possibility of all experience, and assume space as something which lies at the foundation of things in themselves.

397. "But in fact independent of space, there is no other representation, subjective and referring to something external, which could be termed objective *à priori*. For we cannot deduce from any of them synthetical propositions *à priori*, in the same way as from intuitions in space. (3.) Consequently, to speak strictly, no ideality belongs to them, although they accord in this respect with the representation of space, that they belong merely to the subjective property of a mode of sense, as for example, seeing, hearing, feeling, by means of the sensations of colors, sounds, and heat, but which, since they are simply sensations and not intuitions, do not give any object to be known in itself, at least *à priori*.

398. "The object of this observation only goes as far as this,—to prevent us from thinking to explain the asserted ideality of space from extremely insufficient examples: since, namely, perhaps colors, taste, &c., with propriety may be considered not as the property of things, but merely as change of our subject, which may be different even in different men. For in such a case, that which itself originally is only phenomenon, as for example a rose, is held to be valid in the empirical sense, as a thing in itself, which, nevertheless, to each eye, in respect of the color, may appear different. On the contrary, the transcendental conception of phenomena in space is a critical reminding, that nothing generally which is envisaged in space is a thing in itself,—that space is not a form of things which perhaps was proper to them in themselves; but that objects in themselves are not at all known to us, and that what we term external objects, are nothing else but mere representations of our sensibility, whose form is space, but whose true correlative, that is to say, the thing in itself, is not thereby known, and cannot be, but in respect of which also neither is inquiry ever made in experience."*

399. Kant expressly admits "the empirical reality" of both space and time, or their

* Kant, p. 23 *et seq.*

"objective validity in respect of all objects that may ever be offered to our senses. And as our intuition is always sensible, an object can never thus be given to us in experience, which could not stand under the condition of time. On the other hand, we deny to time [and space] all claim to *absolute Reality*, that is to say, that without regard to the form of our sensible intuition, it absolutely inheres in things as condition or property. Such properties as belong to things in themselves, can never be given to us by the senses."*

400. Space and Time are thus merely excluded, like all else of which we take cognizance by our senses or other faculties, from the realm of "absolute reality," while it is admitted that they have the same relative or "empirical reality" as all other objects of which we can acquire any experience or knowledge. If there is no absolute [independent] reality in space and time, there can be no such thing as motion or change. Kant concedes this, "but," he says, "if I could envisage myself, or if any other being could envisage me, without this condition of sensibility, the self-same determinations which we represent to ourselves now, as changes, would then afford us a cognition, in which the representation of time, and consequently also of change, would not at all occur."†

401. It is undoubtedly true, that we do not fully comprehend all the properties, and consequently, all the reality of most of the objects that fall under our cognizance. Intelligent beings might perhaps be differently constituted, so as to "envisage" things under other conditions, and thus to discern a different set of properties, which would convey an idea of reality either more or less adequate than our own. But so vague a hypothesis furnishes no grounds for philosophizing.

402. Everything that has properties either inherent or relative, is a real thing. If the properties are such as belong exclusively to Intelligence, the reality is subjective,—if they belong either wholly or in part to anything else, the reality is objective. Every objective property has, indeed, two sides, one objective, as it exists in the object cognized, and one subjective, as it affects the cognizing mind, and it is in many cases impossible for us to determine the degree of resemblance between the two, or the adequacy of our ideas. But this fact, instead of weakening our belief in objective reality, should rather tend to strengthen the conviction, that every subjective impression is evidence of an objective impress, and that every "empirical reality" that we discover is a representation and evidence, more or less complete, of the true reality.

403. Under this conviction, we may readily assent to Kant's lemma, that "the simple but empirically determined consciousness of my own existence, proves the existence of objects in space out of me," and we may agree with him in rejecting both "the proble-

* Kant, p. 33.

† Kant, p. 34.

matical idealism of Descartes," and "the dogmatical idealism of Berkeley,"* as exclusive systems of philosophy.

404. The following remarks of Derodon† present the objective view of space, very concisely and very happily, although the definitions are naturally, as objective, mostly negative.

"1. Space is not pure nothing, for nothing has no capacity; but *space* has the capacity of receiving body.

"2. It is not an *ens rationis*, for it was occupied by heaven and earth before the birth of man.

"3. It is not an accident inhering in a subject, *i. e.*, body, for body changes its place, but *space* is not moved with it.

"4. It is not the superficies of one body surrounding another, because superficies is an accident; and as superficies is a quantity, it should occupy *space*; but *space* cannot occupy *space*. Besides, the remotest heaven occupies *space*, and has no superficies surrounding it.

"5. It is not the relation or order with reference to certain fixed points, as east, west, north, and south. For if the whole world were round, bodies would change place, and not their order, or they may change their order and not their place, if the sky, with the fixed points, were moved by itself.

"6 and 7. It is not body, nor spirit.

"8. It may be said with probability, that *space* cannot be distinguished from the divine immensity, and therefore from God. It is infinite and eternal, which God only is. He is the place of all being, for no being is out of Him. And although different beings are in different places externally, they are all virtually in the divine immensity."

405. In our examination of the Kantian antinomies, an allusion was incidentally made to Hamilton's belief, that we cannot conceive the possibility either of the finitude or of the infinity of space or time. He says, "We are altogether unable to conceive space as bounded,—as finite; that is, as a whole beyond which there is no further space. Every one is conscious that this is impossible. It contradicts also the supposition of space as a necessary notion; for if we could imagine space as a terminated sphere, and that sphere not itself inclosed in a surrounding space, we should not be obliged to think everything in space; and on the contrary, if we did imagine this terminated sphere as itself in space, in that case we should not have actually conceived all space as a bounded whole. The

* Kant, p. 133–4. Kant's views of space, time, and motion, appear to be nearly the same as those of the Eleatic school. See Anderson, Part III, § 1, and Aristotle, *φυσικῆς ἀκροάσεως*, Book VI, chap. 9.

† Quoted by *Fleming*; Article, SPACE.

one contradictory is thus found inconceivable; we cannot conceive space as positively limited.

406. "On the other hand, we are equally powerless to realize in thought the possibility of the opposite contradictory; we cannot conceive space as infinite, as without limits. You may launch out in thought beyond the solar walk, you may transcend in fancy even the universe of matter, and rise from sphere to sphere in the region of empty space, until imagination sinks exhausted;—with all this, what have you done? You have never gone beyond the finite, you have attained at best only to the indefinite, and the indefinite, however expanded, is still always the finite. . . . Now then, both contradictories are equally inconceivable, and could we limit our attention to one alone, we should deem it at once impossible and absurd, and suppose its unknown opposite as necessarily true. But as we not only can, but are constrained to consider both, we find that both are equally incomprehensible; and yet, though unable to view either as possible, we are forced by a higher law to admit that one, but one only, is necessary. . . .

407. "If we attempt to comprehend time, either in whole or in part, we find that thought is hedged in between two incomprehensibles. . . We are altogether unable to conceive time as commencing; we can easily represent to ourselves time under any relative limitation of commencement and termination, but we are conscious to ourselves of nothing more clearly, than that it would be equally possible to think without thought, as to construe to the mind an absolute commencement, or an absolute termination of time, that is, a beginning and an end, beyond which time is conceived as non-existent. . . We cannot conceive the infinite regress of time; for such a notion could only be realized by the infinite addition in thought of finite times, and such an addition would itself require an eternity for its accomplishment. . . The negation of a commencement of time involves, likewise, the affirmation, that an infinite time has, at every moment, already run; that is, it implies the contradiction, that an infinite has been completed. For the same reasons, we are unable to conceive an infinite progress of time; while the infinite regress and the infinite progress taken together, involve the triple contradiction of an infinite concluded, of an infinite commencing, and of two infinites, not exclusive of each other."*

408. The fundamental difficulty in the foregoing arguments, appears to arise from the ambiguity of the terms *conceive* and *infinite.* If by conception is meant a complete and adequate realization in thought, of infinite space and time, they are undoubtedly inconceivable. But if we use the word only to denote such a degree of knowledge as will enable us positively to assert that space has no bounds, and that eternity has neither beginning nor end, the conception is certainly possible. The "triple contradiction of an

* Hamilton: *Metaphysics,* pp. 527–9.

infinite concluded, of an infinite commencing, and of two infinites, not exclusive of each other," is no contradiction, if we consider that each of the infinites is relative, and that if due regard is paid to their relations, they are in no respect antagonistic or contradictory.

409. No one can set up his own conceptions as an infallible standard for others, but each may contribute to the common treasury of knowledge, his own perceptions of truth. Among the clearest of those perceptions in my mind,—as evident as the simplest axioms of mathematics,—are the infinite extent of space in all possible directions, the infinite duration of eternity, without beginning and without end, and the infinite possibility of position, both in space and time. No truth, no necessary idea has ever been adduced to contradict these conceptions, but the contrary suppositions,—that space is bounded, that duration is transitory, and that position is limited,—lead to countless contradictions and absurdities.

410. We will, therefore, assume as sufficiently established, the three great necessary and infinite Forms,—forms of things, as well as forms of thought,—that correspond to our three forms of Intelligence, and are therefore the only necessary realities of which we can frame any conception. We might, perhaps, imagine a universe, in which there should have been no Being, either material or spiritual, but even in such a universe, in Eternal Silence, Space, Time, and Position would still remain, stern and immovable, ready to be cognized if there were only an intelligence to perceive them.

411. "Space, as containing all things, was by Philo and others, identified with the Infinite. And the text (Acts xvii, 28), which says that 'in God we live, and move, and have our being,' was interpreted to mean that space is an affection or property of the Deity. Sir Isaac Newton maintained that God by existing constitutes time and space. . .

412. "As space is a necessary conception of the human mind, as it is conceived of as infinite, and as an infinite quality, Dr. Clarke thought that from these views, we may argue the existence of an infinite substance, to which this quality belongs." He "maintained that space is an attribute or property of the Infinite Deity."* These opinions, as well as similar views with regard to time and position, find a weighty support in the apparently identical character of the objective properties and our subjective ideas of the three essential forms. If that identity is so marked in our finite intelligence, as to lead many to deny them any other than a subjective reality, it seems probable that the Infinite Intelligence may perceive them merely as attributes of His own Infinite and Eternal Being. Our perception of the necessary existence of the attributes, would then be an evidence of the equally necessary existence of the Being to whom the attributes belong.

* Fleming, p. 481.

CHAPTER XIII.

OBJECTIVE ANALYSIS.

413. In commencing the objective analysis with Essential Necessity, or Form, and proceeding to Essential Reality and Possibility, the method that was pursued in the primary analysis of Consciousness is reversed, and it is therefore well to observe the difference between the chronological order, and the logical order of cognition.

414. "Two ideas being given, we may inquire whether the one does not *suppose* the other; whether the one being admitted, we must not admit the other likewise. This is the *logical* order of ideas. . .

415. "There is still another, that of anterior, or posterior, the order of the relative development of ideas in time,—their *chronological* order. . . Now the idea of space, we have just seen, is clearly the *logical* condition of all sensible experience. Is it also the *chronological* condition of all experience, and of the idea of body? I believe no such thing. . . . Indeed, it is so little true, that the idea of space chronologically supposes the idea of body, that in fact, if you had not the idea of body, you would never have the idea of space."*

416. We have already seen that the ideas of Motivity are obtained the earliest, and those of Rationality the latest. In the order of time, therefore, we proceed from Motivity to Spontaneity, and from Spontaneity to Rationality.

417. But as all truth is based on the absolute or necessary, in the investigation of truth, or the logical order, we descend from Rationality to Spontaneity and Motivity. The chronological order must be first pursued to determine the rational basis, and the logical order subsequently adopted, to erect the superstructure.

418. Substance, in the objective phase of the subjective-objective relation, is the substantial reality that corresponds with the formal reality of space,—or Matter. The intrinsic attributes of Matter, are extension, impenetrability, and shape, all of which are possible only in and through space.

419. Substance in the objective-objective relation, can evidently only be known through analogy. But, inasmuch as various philosophers have been led by various paths, to pronounce primary existence identical with thought or volition, and as it appeared evident at the very outset of our inquiries, that the highest unity, in which the subjective and objective were both united and reconciled, must be a self-cognizing intelligence, we can find

* Cousin: *El. of Psychology*, pp. 85–8.

no representative for this relation but MIND, thus diametrically opposing to Consciousness (the purely subjective upon its own plane of being), that unknown somewhat in which it resides, Mind (the purely objective on its plane of being). The intrinsic attribute of Mind is variety of thought, which is rendered possible only in and through the formal reality of time.

420. Substance in the objective phase of the objective-subjective relation, is the substantial reality that corresponds with the formal reality of position, or FORCE. The intrinsic attributes of Force, are precise limitation, control, direction, all of which are possible only in and through the formal reality of position.

421. Essential Possibility, or Condition, in the objective phase of the subjective-objective relation, is limited, at least so far as we could be able to cognize it, to those displays of being which are in harmony with the formal reality of space, and the substantial reality of matter. Such displays are called phenomenal,—and the primary, most obvious phase of Condition, is therefore PHENOMENON.

422. Condition in the objective-objective relation, embraces those displays of essential being which harmonize with the formal reality of time, and the substantial reality of mind. Mind becomes active and conscious, through relation; relation can be cognized only through difference and succession, in time. The central phase of Condition may therefore be called RELATION.

423. Condition, in the objective phase of the objective-subjective relation, is the Essential Possibility which has a relative harmony with formal position and substantive force. In defining Position, I took occasion to point out the etymological affinity of *place* and *law*. A somewhat similar ideal affinity exists between *force* and *law*, and I therefore designate Condition under its outgoing relation, by the term LAW.*

424. The analysis of Essence is now completed, through the first and second planes of its subdivisions. But Essence itself, is only one of the modes of Existence,—the central and most objective mode, and therefore the one best fitted for testing the possibility of an objective analysis. Taking the analogy of its own primary subdivision, Essence may be regarded as central and quasi-substantial Existence,† toward which flows an incoming Being, analogous to Form,—and from which proceeds an outgoing Being, analogous to Condition. For Existence under the former, or subjective-objective relation, I propose the name of PRINCIPLE,—for objective-subjective Existence, the name of IDEA.

425. Principle, Essence, and Idea, are therefore the three primary classes of Existence, —embracing all things that are the proper objects of finite cognition,—all things that are

* "Laws are the necessary relations which spring from the nature of things." Cousin: *Hist. of Mod. Phil.*, Vol. I, p. 72.

† The Hegelian "Naught."

by their nature placed in relation to our finite intelligence,—all things that are included in the field of objective human philosophy. Whether cognition is possible under other relations, by beings differently constituted from ourselves, we have no means of determining; we may, however, safely assert, that there are no relations inconsistent with those that have been pointed out, and that there can be nothing real or possible, which some necessity does not underlie, and which is not dependent on that necessity for its existence.

426. In Principle, Essence, and Idea (the subjective-objective, objective-objective, and objective-subjective), the answers must be sought to the three great philosophical questions, How?—What?—Why? That which is logically first, is chronologically last, so that though the objective order of creation and dependence may commence with Principle, the subjective order of investigation would proceed from Idea, through Essence, to Principle. Principle and Idea are used synonymously by many writers, but as they are made to include existence under both subjective-objective and objective-subjective relations, it is desirable that the difference of relations should be indicated by different names.

427. If we adopt as objective symbols, P for Principle, E for Essence, and I for Idea, the subdivisions of the next lower plane may be thus represented:

PP, Capacity.	EP, Form.	IP, Quantity.
PE, Subsistence.	EE, Substance.	IE, Quality.
PI, Predication.	EI, Condition.	II, Modality.

428. The relations that are represented by the symbols, are unchangeable, but the names are merely proposed for consideration. If a careful study of the several relations, and a precise determination of their boundaries, shall lead to the suggestion of more fitting terms, they can readily be adopted. The same remarks may be made with regard to the following schedule of the third order of subdivisions:

PPP, Mutability.	PPE, Tendency.	PPI, Consequence.
PEP, Ordination.	PEE, Efficiency.	PEI, Dependence.
PIP, Accident.	PIE, Species.	PII, Genus.
EPP, Space.	EPE, Time.	EPI, Position.
EEP, Matter.	EEE, Mind.	EEI, Force.
EIP, Phenomenon.	EIE, Relation.	EII, Law.
IPP, Unity.	IPE, Plurality.	IPI, Totality.
IEP, Negation.	IEE, Limitation.	IEI, Affirmation.
IIP, Necessity.	IIE, Reality.	III, Possibility.

429. The following classification, in accordance with symbolic resemblances, will perhaps render the meaning of the several terms clearer and more definite.

1. Class of pure Principle.

PPP, Mutability.

2. Class of duplicate Principle, and Essence.

PPE, Tendency.	Essential Capacity (PP, E); subsistent Principle (P, PE).
PEP, Ordination.	Principal Subsistence (PE, P); formal Principle (P, EP).
EPP, Space.	Principal Form (EP, P); capable Essence (E, PP).

3. Class of duplicate Principle, and Idea.

PPI, Consequence.	Ideal Capacity (PP, I); predicative Principle (P, PI).
PIP, Accident.	Principal Predication (PI, P); quantitative Principle (P, IP).
IPP, Unity.	Principal Quantity (IP, P); capable Idea (I, PP).

4. Class of Principle, and duplicate Essence.

PEE, Efficiency.	Essential Subsistence (PE, E); substantial Principle (P, EE).
EPE, Time.	Essential Form (EP, E); subsistent Essence (E, PE).
EEP, Matter.	Principal Substance (EE, P); formal Essence (E, EP).

5. Class of Principle, Essence, and Idea.

PEI, Dependence.	Ideal Subsistence (PE, I); conditional Principle (P, EI).
PIE, Species.	Essential Predication (PI, E); qualitative Principle (P, IE).
EPI, Position.	Ideal Form (EP, I); predicative Essence (E, PI).
EIP, Phenomenon.	Principal Condition (EI, P); quantitative Essence (E, IP).
IPE, Plurality.	Essential Quantity (IP, E); subsistent Idea (I, PE).
IEP, Negation.	Principal Quality (IE, P); formal Idea (I, EP).

6. Class of Principle, and duplicate Idea.

PII, Genus.	Ideal Predication (PI, I); modal Principle (P, II).
IPI, Totality.	Ideal Quantity (IP, I); predicative Idea (I, PI).
IIP, Necessity.	Principal Modality (II, P); quantitative Idea (I, IP).

7. Class of pure Essence.

EEE, Mind.

8. Class of duplicate Essence, and Idea.

EEI, Force.	Ideal Substance (EE, I); conditional Essence (E, EI)
EIE, Relation.	Essential Condition (EI, E); qualitative Essence (E, IE).
IEE, Limitation.	Essential Quality (IE, E); substantive Idea (I, EE).

9. Class of Essence, and duplicate Idea.

EII, Law.	Ideal Condition (EI, I); modal Essence (E, II).
IEI, Affirmation.	Ideal Quality (IE, I); conditional Idea (I, EI).
IIE, Reality.	Essential Modality (II, E); qualitative Idea (I, IE).

10. Class of pure Idea.

III, Possibility.

430. This schedule of the objective has been ideally or subjectively determined, and the objective validity of the determination may perhaps still be doubted. Because all our knowledge is obtained through finite ideal representations of real objects, we are apt to think that such is the order of nature, and to regard the ideal as necessarily and universally dependent upon the real.

431. But even in finite intelligence, we may find reasons for reversing or modifying that opinion, for in every new invention that springs from the limited creative power which is typical of the Infinite creative power of an Infinite Intelligence, the idea is evidently antecedent and constitutive. The realized steam-engine is the adequate representative of all the ideas of its successive inventors and improvers, but our mental notion of the engine is generally an inadequate *re*-representation of the adequate representation. Our ideas, therefore, so far as they are representative, are but secondarily so, while the objective universe is primarily and immediately representative of the originally presentative idea of the Supreme Architect,—who has ideally determined the forms and relations of all reality.

432. No place has been assigned in the objective schema, for the Consciousness on which all our analysis is based. In one of its phases, it may be regarded as subordinate to the purely objective Mind (EEE), and as constituting one of its subdivisions. But Consciousness is authoritative, not merely on the plane of mind. Whether we ascend or descend,—whether we contemplate the highest conceivable forms of being, or its lowest and minutest subdivisions, the plane of Consciousness shifts with our shifting points of view. The pure subjective is the image of the Divine, inasmuch as it stands in correlation to everything, maintaining throughout a distinct independence, which entitles it to the place we at first assigned it, as one of the forms of the highest duality which is immediately subordinated to the highest Unity.

433. The accompanying diagram exhibits the relations, together with the experimental nomenclature, of the several subdivisions of objective Being. In the chart of Consciousness, as each of the subdivisions is still fundamentally subjective, and as each successive analysis is based on the same primary relations, the symbolic order of M, S, R, is retained throughout. But in studying the Objective, we attempt to penetrate the mysteries of Being as it is in itself, independent of subjective relations, and it therefore seems proper to adopt an arrangement which will group the several subdivisions more systematically around the central symbols. This variety of possible collocation under uniformity of law, is an instance of the multiform *tendency* to system prevailing in nature, which is much more agreeable than absolute symmetry, and more consonant with the theory of a Supreme Free Intelligence. Other instances of the harmonious blending of Liberty and Law, may be found in Phyllotaxis,—in the different classifications of Natural Science,—and indeed,

Intellectual Symbolism by Pliny E. Chase M.A.

ELEMENTARY FORM
UNDER VARI
PIE
PII
PIP
SPECIES
GENUS
ACCIDENT
PI
PPI
CONSEQUENCE
PPP
MUTABILITY
PP
PPE
TENDENCY
CAPACITY
PREDICATION
P
Principle
OBJE
PEP
ORDINATION
SUBSISTENCE
PEE
EFFICIENCY
PE
DEPENDENCE
PEI
FORM
POSITION
EP
EPI
SPACE
TIME
MATTER
EPP
EPE
EEP

P. S. Duval & Son Lith. Philad.

in almost every attempt of philosophy to interpret the various developments of Divine Idea.

434. If we suppose Existence and Intelligibility* to be convertible terms, the three momenta of the Intelligence which sustains the Intelligible,—or Intelligence affected, Intelligence *per se*, and Intelligence affecting,—may perhaps be represented by Consciousness, Power, and Manifestation. As Existence is transmitted through all time and space, in infinite ramifications, we may readily suppose it to be accompanied by the parallel ramifications of Intelligence, or even to be identical with them. The general relations of the subjective and objective, which we have attempted to trace by a symbolic analysis, might thus be infinitely modified, and our deductions would become like mathematical formulas, which require a due regard to the conditions of every problem to which they are applied, in order to determine their concrete significance.

435. We have as yet but few data for judging of the adequacy of such a hypothesis, or of the effect that would be produced by extending the system which we have briefly examined under our elementary Intellectual development, so as to embrace the Moral and the Practical, the Personal and the Social, with their myriad forms and groups of new combinations and relations. But if we give a moment's reflection to the variety of those possible relations, the need of a good system of classification will be evident, and the great impulse that was given to mathematical discovery by the broad generalizations of the Calculus, shows the advantage of combining with such a system a language of symbols that will concisely, and at the same time plainly, embody the results of long and patient study, for future convenient use and reference.†

436. The very errors of speculation, seem to confirm the fundamental relations of our present analysis. If it be admitted that every doctrine must have some basis of truth, in order to commend itself to any intelligent acceptance, the erroneous exaggerations which give undue prominence to favorite views, may be expected to show a natural grouping about certain primitive points of relationship. Such a grouping is evident in Cousin's division of philosophic systems into Sensualism, Skepticism, Idealism, and Mysticism.

437. If we place a too exclusive dependence on the affection of Consciousness by the

* Not the imperfect Intelligibility of a finite mind, but that of the Infinite and All-pervading.

† The ternary division is not *necessarily* applicable, except when we wish to represent the modifications of any single power under its relations to itself and to another power. The successive subdivisions according to the same law are, however, perfectly natural, and may be extended as far as may seem desirable, in order to accomplish any special intellectual purpose. Whenever any such purpose can be better accomplished by a binary, quaternary, or other division, such a division should undoubtedly be adopted, but even then, a close examination would perhaps enable us to trace the new basis of classification to some blending and modification of two or more relative triplicities.

objects of its cognition (OS), the senses, which are the immediate avenues for our intercourse with the external world, may be regarded as the sole instruments of knowledge, and the sensual school will then appear to embrace all the correct expounders of the mysteries of nature. If the mind be assumed as its own sole interpreter of truth (SS), and no importance is attached to any external means for verifying its inferences or removing whatever doubts may arise, we shall soon be landed in inevitable Skepticism. If the reality of things be regarded as precisely commensurate with the reality of thought (SO), the broad foundation of Idealism is laid. If an attempt be made to exclude all mental bias and coloring, and to penetrate the crust of phenomena, in order to ascertain the substantial nature of things as they are in themselves (OO), the unconscious deductions of analogy may be assumed as the illuminations of absolute, unmodified, and unrelated truth, and the result will be Mysticism.

438. To a perfect intelligence, it might be a matter of indifference what system of exegesis should be adopted, but the finite mind cannot concentrate its attention upon any single point, without slighting to some extent, other points that are important in forming accurate general conclusions. A liberal Eclecticism, that attempts to embrace in its range of vision the whole landscape of truth, may have the most correct idea of the relative bearing of all the different portions, but it will lose many of the most beautiful features that a closer local inspection would discover. Although its creed may contain the most correct exposition of the "Common Sense" of the race at the moment, it will contribute comparatively little, except by its exposition of the true state and needs of philosophy, to the progress that is mainly effected by myriads of co-workers, who may each be men of "one idea," but whose combined labors tend to the contemporaneous development of many ideas.

CHAPTER XIV.

THE ABSOLUTE.

439. The goal, as well as the starting-point,—the Omega as well as the Alpha of Philosophy, is the Absolute.* The love of wisdom commences in Faith, and in Faith alone can

* "There are three terms, familiar as household words, in the vocabulary of philosophy, which must be taken into account in every system of Metaphysical Theology. To conceive the Deity as He is, we must conceive Him as First Cause, as Absolute, and as Infinite. By the *First Cause*, is meant that which produces all things, and is itself produced of none. By the *Absolute*, is meant that which exists in and by itself, having no necessary rela-

it find any final resting-place. The sages of India and China, of Persia and Egypt, of Greece and Rome, have all sought in vain to find out by unaided reason, the Infinite and Eternal One, who is "without variableness or shadow of turning." By successively rejecting all the perceived relations which are felt to be the restraints or limits of imperfection, they have arrived at a dim shadowy Idea, which is supposed to be devoid of all relation, and of which nothing can be predicated except negations and a name.

440. What is this ghostly Idea,—this idol of the finite Intelligence, but the highest or most abstract form of that inappreciable, and so far as the human Consciousness is concerned, non-existent relation, which we have designated as the Objective-Objective.* Reason, conscious of her own weakness, and yet confident in her instinctive belief that there must be some Greatest, regards the Objective under relation to herself, whether that relation be Objective-Subjective or Subjective-Objective, as limited by the relation, and therefore, imperfect and finite. But in the Objective-Objective, if there be any limitation, it can never be appreciable by human intelligence, and there, if anywhere, must exist that abstract, underlying Infinite, which is in itself devoid of all relation. The belief on which this deduction is based, so far as it is instinctive, is one of those primary revelations of faith, which is infallible, provided it is received in its primitive simplicity, and whatever error may be supposed to spring from it, must arise from the gloss of imperfect human apprehension. Let us examine a few of the formulas that have been devised for the expression of this almost universal creed of humanity, in order to ascertain, if possible, the precise extent of the truth that it represents.

441. Anaximander. "The original essence which he assumed, and which he is said to have been the first to have named principle (ἀρχή), he defined as the 'unlimited, eternal, and unconditioned,' as that which embraced all things, and ruled all things, and which, since it lay at the basis of all determinations of the finite and the changeable, is itself infinite and undeterminate."†

442. Aristotle. "Hence the famed Aristotelian definition of the Absolute, that it is

tion to any other Being. By the *Infinite,* is meant that which is free from all possible limitation; that than which a greater is inconceivable; and which, consequently, can receive no additional attribute or mode of existence, which it had not from all eternity." *Mansel,* p. 75.

* "God, considered without relation with the world and humanity, undoubtedly still exists. He exists wholly in the depths of His essence, invisible, inaccessible, incomprehensible; but this is no longer the God of the world and the God of humanity; it is no longer a God who overlooks and superintends His work, the God whom men adore and bless under the name of Providence." *Cousin: Hist. of Mod. Phil.,* Vol. I, p. 163. "The Bhagavad-Gita expressly teaches that, in the hierarchy of the human faculties, the soul is above sensibility, that above the soul is intelligence, and that there is something still above intelligence,—being." *Ib.,* Vol. I, pp. 392-3.

† Schwegler, p. 22.

the thought of thought (νόησις νοήσεως), the personal unity of the thinking and the thought, of the knowing and the known, the absolute subject-object. In the Metaphysics (XII, 1) we have a statement in order of these attributes of the Divine Spirit, and an almost devout sketch of the eternally blessed Deity, knowing Himself in His eternal tranquillity as the absolute truth, satisfied with Himself, and wanting neither in activity nor in any virtue."*

443. Herbart. "If the world actually exists as a whole, disposed according to design, it follows that we must inquire also for the author of this arrangement, and shall find him in an essence that is above us, but not merely within *our* vision, which would only transfer the reason of man to nature. This belief in a Spirit of Order, little as it is grounded on demonstration, yet depends directly upon the same conclusions, and has the same certainty as the belief by which every man is convinced of the existence of other rational spirits; for of my fellow-men I see only forms and teleological acts, and that these proceed from rational thought is only a belief, but one so worthy of confidence that it stands in certainty above all knowledge."†

444. Schelling. "To bring Realism and Idealism into a state of reciprocal penetration, such has been the declared object of all my endeavors. The notion of the absolute substance, ... obtained by the higher method of contemplating nature, and from the unity that was recognized as subsisting between the dynamical and the psychical or mental, a living basis, out of which grew the Philosophy of Nature, ... which, when considered in reference to the whole of philosophy, must invariably be regarded as that real portion of the latter, which, by a process of redintegration through the influence of the ideal, in which freedom prevails, becomes susceptible of elevation into the true sphere or system of rational thought."‡

445. "'Our mind strives after unity in the system of its knowledge; it will not endure that there should be pressed upon it a separate principle for every single phenomenon, and it will only believe that it sees nature when it can discover the greatest simplicity of laws in the greatest multiplicity of phenomena, and the highest frugality of means in the highest prodigality of effects. Therefore, every thought, even that which is now rough and crude, merits attention so soon as it tends towards the simplifying of principles, and if it serves no other end, it at least strengthens the impulse to investigate and trace out the hidden process of nature.' The special tendency of the scientific investigation of nature which prevailed at that time, was to make a duality of forces the predominant element in the life of nature. . . . . In opposition to these dualities, Schelling now insisted upon the unity of everything opposite, the unity of all dualities, and this not simply as

* Schwegler, p. 126. † Chalybäus, p. 136. ‡ Ib. p. 265.

an abstract unity, but as a concrete identity, as the harmonious co-working of the heterogeneous."*

446. "How is it possible that our thought should ever rule over the world of sense, if the representation is conditioned in its origin by the objective? The solution of this problem, which is the highest of transcendental philosophy, is the answer to the question: how can the representations be conceived as directing themselves according to the objects, and at the same time the objects conceived as directing themselves according to the representations? This is only conceivable on the ground that the activity through which the objective world is produced, is originally identical with that which utters itself in the will."†

447. HEGEL. "Now, those counterparts, or opposites, bear reference altogether to the definite antagonism of indifference and difference, identity and difference, matter and form, internal and external, and especially positive and negative. It is true that by essence we usually think at first of the substratum, which has in itself certain determinate states, or which lies at the bottom of these. These determinate states, modes, and forms, ... are not to be separated from the essence, but to every present appearance there must be at bottom a real or essential element, or, to use the expression of Herbart, for every appearance there must be a real to which the former points."‡

448. "We thus arrive as a result at the Aristotelian νόησις τῆς νοήσεως, the self-thinking process of thought, or the self-knowing truth, Absolute Idealism, which in itself is absolute realism, or an identity in which those antagonisms have coalesced, in order to generate or engender themselves anew, without positing therewith a duplicity of principles; seeing that the production of the antagonisms, or, to speak more concisely, the powers of self-opposition or *absolute negativity*, is the one absolutely self-moving principle."§

449. Hamilton, as we have seen, uses the term Absolute to denote what is "*finished, perfected, completed;* in which sense the Absolute will be what is out of relation, &c., as finished, perfect, complete, total." "In this acceptation . . . the Absolute is diametrically opposed to, is contradictory of, the Infinite." He also speaks of the Unconditioned as "the genus of which the Infinite and Absolute are species," and says that "the Absolute and Infinite are conceived only as negations of the conditioned in its opposite poles."‖ We have already examined some of the contradictions in which, by his own admission, these definitions necessarily involved their author, and those necessary contradictions might be reasonably assumed as sufficient evidence of error. But even without regard to the consequences of the definitions, what can be "the opposite pole" to the Infinite or

* Schwegler, p. 317.
† Ib. p. 323.
‡ Chalybäus, p. 325.
§ Ib. p. 311.
‖ *Discussions*, pp. 21, 36.

Unlimited if it be not the Finite or Limited? The limited must be conditioned by its limits,—therefore the unconditionally limited is an absurdity. On the other hand, the unconditioned is necessarily unlimited, or infinite.

450. To the doctrine of Cousin, that the idea of the infinite, or absolute, and the idea of the finite, or relative, are equally real, because the notion of the one necessarily suggests the notion of the other, Hamilton replies:

451. "Correlations certainly suggest each other, but correlations may, or may not be equally real and positive. In thought, contradictories necessarily imply each other, for the knowledge of contradictories is one. But the reality of one contradictory, so far from guaranteeing the reality of the other, is nothing else than its negation. . . . . It therefore behooved M. Cousin, instead of assuming the objective correality of his two elements on the fact of their subjective correlation, to have suspected, on this very ground, that the reality of the one was inconsistent with the reality of the other."*

452. No one ever claimed that the finite mind could fully comprehend or understand the Infinite, but we certainly have so far the power of conceiving it, as to positively assert its existence. If of the two correlations, the finite and infinite, either is unknown, is it not the finite? The infinite is wholly independent of any subjective relation or coloring, but the finite is apprehended only under the subjective relations which we assign it, and it may plausibly be regarded as merely a subjective notion, destitute of any objective reality. The Infinite and Unconditioned are certainly objects of thought, and though we may know nothing more *of* them (as we know nothing more of mind and matter), than their relative manifestations, we may know that there is something more *in* them. Though we cannot identify ourselves with the Absolute in reality, may we not cognize it under ideal relations? When we think of our own subjectivity, how is the "subject contradistinguished from the object of thought,"† except in idea?

453. A recent writer on "the philosophy of the Infinite," thus notices Hamilton's argument,—that if, in any instance, we imagine that we obtain a knowledge of the Infinite, we only deceive ourselves by substituting *the indefinite* for the infinite.‡

454. "While we endeavor to answer this argument, let it be remembered that both Sir William and we have this common ground,—that the Indefinite is only a characteristic

* *Discussions*, p. 34.

† "The mind knows nothing, except in parts, by quality, and difference, and relation; consciousness supposes the subject contradistinguished from the object of thought; the abstraction of this contrast is a negation of consciousness; and the negation of consciousness is the annihilation of thought itself." *Discussions*, p. 26.

‡ "Condillac denies the infinite, unity, substance, etc., and reduces everything to the indefinite, to the finite multiplied by itself, to a simple collection of quantities and accidents, etc." *Cousin: Hist. of Mod. Phil.*, Vol. I, p. 178.

of thought; while the Infinite is an object of thought. . . . . We admit to Sir William that the knowledge which we have been describing, and the knowledge of the Infinite, which we intend to describe at still greater length, is an *indefinite knowledge.* But it is an indefinite knowledge of what? Of this: It is an *indefinite knowledge* of an *infinite object.* It is not a knowledge of the finite, for we can find no limits; according to our own consciousness, and according to Sir William's statement, it is an indefinite knowledge of something; therefore it is an indefinite knowledge of the infinite. We profess nothing but an *indefinite* knowledge, but it must be a knowledge of something, and as not of the finite, it must be of the infinite. Sir William's argument we consider valid, if viewed as a refutation of the assertion that we have a clear and definite knowledge of the Infinite. But, on the other hand, Sir William maintains for himself that we can have no knowledge of the Infinite. This conclusion we consider no more valid than the other, for it does not follow that, since we have not a clear and definite knowledge of the Infinite, therefore we can have no knowledge of it at all. . . . . While it is true that the finite mind cannot have infinite thoughts, we hold it equally true that the finite mind can have finite thoughts concerning an infinite object. . . . In so far as Sir William maintains that we cannot have a clear knowledge of the Infinite in all its extent; and in so far as M. Cousin maintains that we can have some knowledge of the Infinite; we consider that they both are right."*

455. A limited idea may be a partial representation of an unlimited reality. "Time is the image of eternity," and in like manner the human Consciousness may be an image, though a faint one, of the Infinite Intelligence. In the variety of possible relations, each single relation may furnish the expression of a partial truth, and the greater the number of relations that is brought under our cognizance, the more nearly adequate will be our conception of the reality.

456. Relation does not necessarily limit anything but conception, and the Absolute, though independent of all relation, may and does place itself in relation to finite intelligence, without detracting from its own infinite perfection. The Infinite might, perhaps, even become self-limited, in certain directions, for the accomplishment of its own purposes; —it certainly does not become us to deny such a possibility, when by making the denial, we attempt to make our own conceptions the limits of Divine power.

457. Among the great controlling forces of the material universe, light, heat, electricity in its various forms, attraction, inertia, repulsion, there are many indications of unity. And although the principles of correct philosophizing will not allow us to assume their essential identity until it is more fully demonstrated, the same natural instinct of Reason that

* Calderwood, pp. 77–79.

has enabled us to approach so nearly to a unity of force, compels us to mount still higher, until we attain to the conception of the Necessary, Supreme Being, who is absolutely knowable, so far as the reality of His existence is concerned, and at the same time absolutely incomprehensible, and only faintly shadowed forth under such relations as He sees fit to employ in His revelation to His intelligent creatures.

458. But while Reason thus ascends through the analogies of Spontaneity to the Greatest, and after the successive abstraction of all subordinate relations, finally abstracts relation itself, and thus forms the conception of the Absolute, there are other needs of our nature that equally demand supreme satisfaction. In all the impulses of Motivity, there is an underlying idea of good, which points to some infallible Best, and in Intelligence, whose especial province it is to know how to accomplish ends, we see a like pointing to some Wisest, whose purposes could not be thwarted by any inferior Intelligence.

459. Following these pointings of its co-ordinate faculties, Rationality discovers that the Absolute must also be the Best, Greatest, and Wisest. The Supreme Being must be self-impelling, or active solely for subjective reasons, otherwise He would be subordinate to a higher power; self-acting, or He would not be Supreme; intelligent, or He could not be self-acting.*

460. Necessity is but another name for accordance with the eternal and unchangeable Divine Will, and so far as necessity becomes evident, we are able to perceive Divine determination. Philosophy, resting on revelation, may use all the facts and teachings of faith as the materials of its deductions, and it should verify its results by comparing them with the well-recognized truths of revelation. By making such a comparison, it will ascertain that there is not only an Absolute, which is in itself entirely independent of all external relation, and therefore entirely aloof from all possible human conception, except

* "Aided by instruments which the necessities of reason itself have called into existence, man, in these last times, has well demonstrated the homogeneousness of his mind with the Supreme Creative Mind, and he has done so on a field not narrow, for it is as wide as the stellar universe. There can be no irreverence,—there can be no presumption in plainly stating a fact which rests upon evidence so clear and sure. Even if this same averment were made in terms still stronger and more comprehensive, we need not fear a rebuke on the part of Christian piety, for what we so affirm does but illustrate and attest the Biblical doctrine, that 'God made man in his own image.'" *Taylor: World of Mind*, p. 167.

"In the prosecution of the modern Physical Sciences, the human mind has demonstrated the congruity of the human Reason with *that* REASON of which the material universe is the product; for when we say that (within certain limits) we understand the scheme of the world as to its structure and as to its dynamics, we affirm that the mind which understands and the MIND which has produced this scheme of things are in unison, or that they are convertible, the one into the other." *Ib.* p. 327.

in the mere idea of Existence, but that there is also an ABSOLUTE-RELATIVE, which has voluntarily placed itself in relation with its works.*

461. Only under the conception of such an Absolute-Relative is it possible for us to believe in a Divine Love that works teleologically for benevolent purposes,—a Divine Activity, "upholding all things by the word of His power,"—a Divine Wisdom by which all things were made,—and thus to form some idea, however feeble it may be, of that ineffable Divine Image in which man was created. Only under such a conception can we understand that in our ascent to the Absolute, we can never overstep that Highest Unity, which is at once Knowing and Known, and that the atheistical notion of a blind fate, an unreasoning power, an unintelligent "order of nature," is opposed to the plainest deductions of reason, as well as to the teachings of revelation. Only under such a conception can we feel that man has not been created in vain,—that all the needs of our toiling, struggling, suffering race, have been provided with the means of satisfaction,—that the Infinite Creator is "not a God afar off," but that He is "Our Father" in heaven, with more than a father's sympathy and love,—"the Lord merciful and gracious, long-suffering, and abundant in goodness and truth."

462. Under this conception, what a halo of glorious beauty is thrown around the Christian Revelation! However imperfect may be our appreciation of the Divine mysteries,—however feeble and halting our faith,—however sad and desponding our hope,—or however arrogant our self-sufficiency and impatience under fancied dictation,—the dignity, simplicity, and majesty portrayed in the inspired records, will be accepted as the manifest indications of that mysterious blending of the Absolute with the Relative,—the Divine with the Human, in which faith and philosophy both find their final resting-place.

463. He who wept at the grave of Lazarus, gave eternal evidence of the intimate union between Divine love and human sympathy with suffering. The charity that, in the midst of the agony in the garden, forgave the weakness of the heavy-eyed disciples, will ever

* "I will repeat my innermost conviction, that the existence of free will, in man as a fact of the consciousness, in God as an object of our faith, is the cardinal point on which all that is sound in philosophy or true in religion ultimately turns. Banish it from heaven and earth, and men become nothing more than the petty wheels in the vast machine, of which God is the involuntary motive power. But the engineer is absent, and he whom we then call God, is bound in the trammels of a merciless necessity,—no object of love, for he cannot hate, no object of prayer, for he cannot aid, no object of praise, for he is a tool in the hands of a higher fate. . . . . Restore free will, and where all before was death and darkness, all now becomes life and light. Then indeed does God cease to be the omnipresent automaton, the dead God of the dead, and becomes a spirit and a power, and the living God of the living. Then, indeed, are love, and prayer, and praise, His just meed and our high privilege, for of His own free will, and of no necessity, is He a gracious God and merciful, slow to anger and of great kindness, and His tender mercies are over all His works. For of Him, and through Him, and to Him, are all things; to whom be glory forever." *Solly*, p. 282.

continue to inspire erring humanity with the hope of forgiveness, for the consequences of its own feebleness and imperfection. The great mystery of all mysteries,—God the Incarnate Word, by whom all things were made, who came unto His own, and His own received Him not, in whom was life, and the life was the true Light which lighteth every man that cometh into the world, is at once a revealed evidence, and a philosophical consequence of the mediating relativity of the Absolute. Here, then, may baffled speculation, which in its endeavors to grasp the unrelated Infinite, has laid a vain oblation on the altar of "the unknown God," at length find satisfaction, and bow reverently and thankfully as it hears the voice of the once persecuting, but afterwards converted and zealous apostle, proclaiming to the philosophers of all time, as well as to the "too-superstitious" Athenians,—"Whom therefore ye ignorantly worship, Him declare I unto you."

SUMMARY.

PREFACE, pages 463–467.

Intellectual endowments, the echo of the Almighty Mind—Connection of religion and philosophy—Natural tendency to inquiry—Faith and Reason are handmaidens—Knowledge rests on revelation—Proper question of faith—Relation, the natural basis of mental classification—Generalization is superficial—Uses of notation—Position, as eternal and necessary as Space and Time—Direction of discovery, indicated by tentative analysis—Imperfection of a first Essay—Skepticism the limit of faith—Absolute Skepticism impossible.

CHAPTER I.—Definitions and Fundamental Relations, pages 467–475.

Science and Faith, § 1—Self-evident beliefs are revelations, 4—Knowledge implies a dual existence, 12—The highest Unity is Intelligent, 14—Triplicity of Mental Faculties, 17—Dictum of Aristotle, 19—Relativity the basis of Hegel's philosophy, 20—And of all philosophy, 22—Limited number of relations, 23—Consciousness essential to Mind, 28—It involves Time, 30—Motivity, Spontaneity, and Rationality, 31—Mutuality of Faculties, 34—Elementary Symbols, 36—Propriety of names, 38—Reid's division, Understanding and Will, 40—Quantitative division of faculties, 42.

CHAPTER II.—Primary Faculties, pages 476–486.

Philosophical recognition of Motivity and Spontaneity, § 43—Propensity, 45—Desire, 49—Sentiment, 53—Instinct, 56—Will, 60—Energy, 67—Rational faculties, adapted to various ways of acquiring Knowledge, 71—Perception, 76—Judgment, 82—Understanding, 93—Progressive development of faculties, 98—Relation of the objective to the subjective incomprehensible, 100—The Divine Reason, 101—Necessary cognitions, native to the mind, 104—Hamilton's six special cognitive faculties, 106—Their gradation from objective to subjective antecedence, 107.

CHAPTER III.—Subordinate Faculties, pages 486–494.

Three guides to nomenclature, § 108—List of secondary faculties, 110—Their assignment, 111—Comparative analysis of special faculties, 115—Homogeneous classification, 122—Questions for determining fitness of names, 123—Tertiary faculties, 124—The first essay tentative; difficulties of classification, 125—Diagram of Elementary forms of Consciousness, 126.

CHAPTER IV.—Philosophy of Consciousness, pages 494–503.

Personality not physical, § 127—Only phenomena can be studied, 128—Mind more intelligible than matter, 128—Essential and accidental attributes, 129—Fichte, Cousin, and Hamilton, 130—Discouragements of superficial philosophy, 134—Advantages of system, 136—Is the Mind always conscious? 137—Steps in Knowledge, 142—Accountability, 147—Control of Motivity, 150—Free Will, 153—Tendency of Spontaneity, 155—And Rationality, 157—Steps to perception, 159—Rapidity of volition, 160—Solid foundation of Psychology, 161.

CHAPTER V.—Knowledge and Faith, pages 504–513.

Object of science, § 163—Demonstration, 164—Difference in character of propositions, 170—Sensual Knowledge the most obvious, 171—Mathematics, the science of form and proportion, 174—Do our senses deceive us? 177—Sources of error illustrated, 178—Mathematical fallacies, 182—Sensations indubitable, 183—Use of judgment in vision, 184—Deception, 186—Revelation, 191—Error is based on truth, 194—Revelation, infallible but partial, 197—Spiritual perception, 198—Culture of faculties, 201—Imperfect development of spiritual nature, 204—Authority, 205—Theorems may become axioms, 206—Demonstrations of Divine Existence, 209—Self-evidence the test of truth, 211.

CHAPTER VI.—Character and Limits of Belief and Certainty, pages 514–522.

Distinction of knowledge and belief, § 212—Demonstrability not the only test of certainty, 214—Metaphysical science possible, 217—The senses must be believed, 218—Facts of consciousness, 220—Apperceptions of reason, 222—Limits and characteristics of knowledge, 227—Differences in extent of knowledge, 229—Axioms discovered slowly, 230—Intuition the basis of knowledge, 233—Knowledge is clear and distinct, 235—Faith is natural, 237—Dogmatism should be avoided, 239—Harmony of faith and reason, 240—Revelation adequate to our needs, 242.

CHAPTER VII.—Rational Antinomies, pages 522–527.

Possible antagonisms of Faith, none of Reason, § 243—Paradoxes from improper use of terms, 244—The Absolute, 247—Defined by Hamilton, 248—Hegelian Zero, 249—Source of contradictions of the Absolute, 250—Kant's skeptical method, 251—His first Antinomy, 252.

CHAPTER VIII.—Examination of Antinomies, pages 527–534.

Ambiguity of the terms "world" and "infinite," § 253—Achilles and the tortoise, 255—Other ambiguities of Kant, 256—Hamilton's "Contradictions," 261—Their leading ambiguity, 262—Mill's criticism of common maxims, 264—His views examined, 268—Modifications of Force, 269—All paradoxes can be solved, 271.

CHAPTER IX.—Basis of Ontology, pages 535–541.

Faith, the ground of objective knowledge, § 272—Analogy, 273—Subjective source of all things, 274—Ontology, or Transcendental Philosophy, 275—Reasons for objective trichotomy, 276—Approaches to a universal lan-

INDEX OF NAMES.*

* Some of my notes were made so long ago, that I may have inadvertently neglected to make due acknowledgments. It is impossible, even with extraordinary care, to distinguish at all times between thoughts that are original, and those that are borrowed.

GENERAL INDEX.

MILANO

TIPOGRAFIA DI GIUSEPPE BERNARDONI

1867.

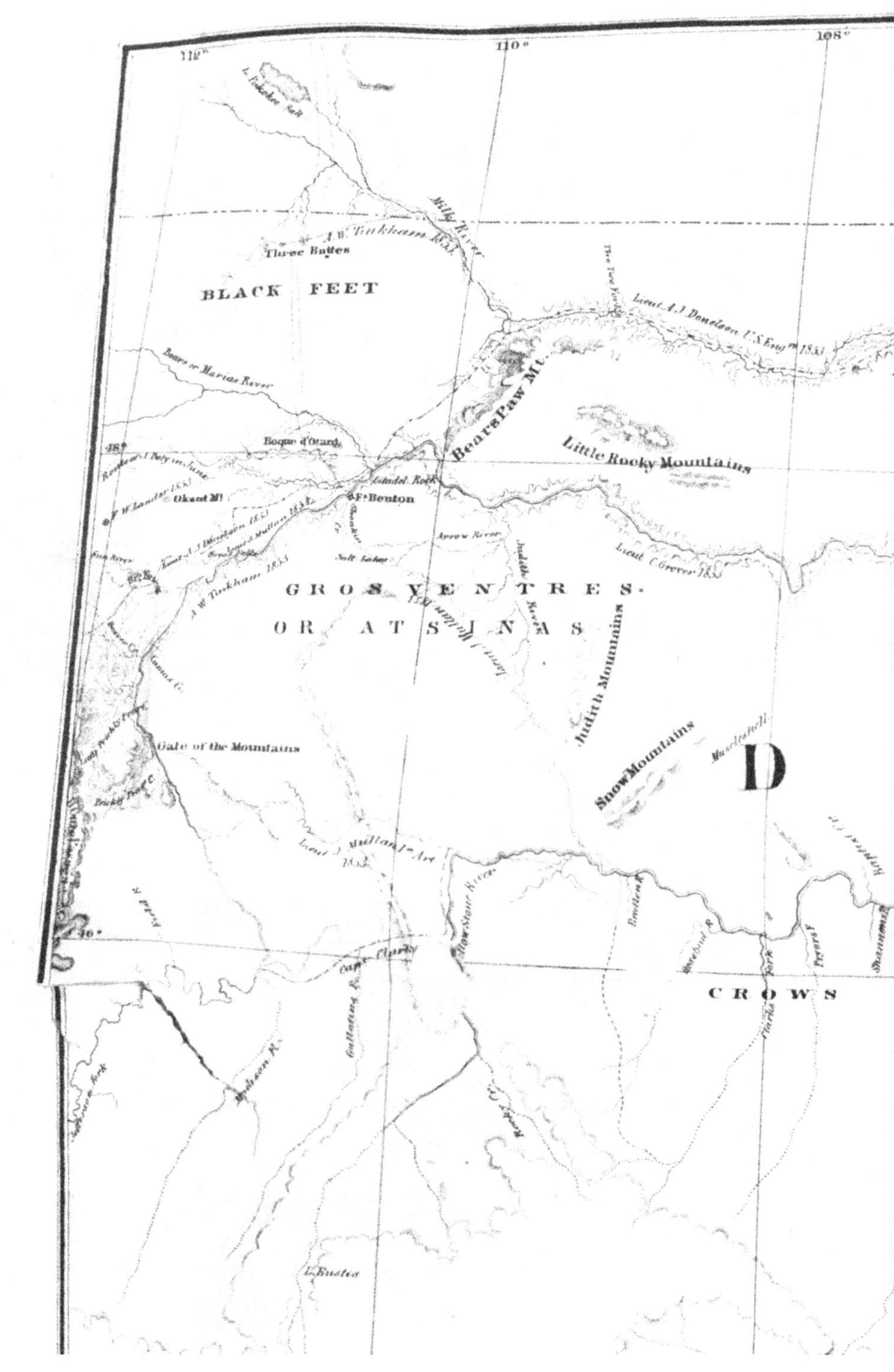

110°
108°
Three Buttes
BLACK FEET
Bears Paw Mt
Little Rocky Mountains
Boque d'Otard
Citadel Rock
Ft. Benton
Okaut Mt
Arrow River
Salt Lakes
Lieut. A. J. Donelson U.S. Eng. 1853
A. W. Tinkham 1853
GROS VENTRES
OR ATSINAS
Judith Mountains
Snow Mountains
Gate of the Mountains
D
Lieut. J. Mullan 1st Art.
1853
Capt. Clark
CROWS
Madison R.
Gallatin R.
L. Eustis

106°
104°
Common Hunting Ground
of
BLACK FEET & ASSINIBOINS
ASSINIBOINS
MINETARES
Panther Hill
RIVER
SOURI
Ft. Union
UNEXPLORED
A
K
O
Capt. Clark 1807
YELLOW STONE RIVER
Ft. Alexander
Big Horn R.
Little Horn R.
Tongue R.
Powder R.
Box Elder R.
BLACK FEET AND UNK
DAKOTAS
UNEXP
MINIKANYES

MINI WAKAN
Maison du Chien
Ft Berthold
Ft Clark
MANDANS
&
ARI CAREES
IHANKTONWANNA
DAKOTAS
T
A
PAPA
Thunder Hill
ONPA DAKOTAS
I
N

94°

92°

48°

MINNESOTA

Rainy Lake

Rainy Lake R.

Little Fork

Big Fork

Turtle L.

Lake Bemidji

Cass L.

Leech L.

L. Itasca

Swan L.

Sandy L.

Fond du Lac

Sand Hill R.

Wild Rice R.

Otter Tail L.

Leaf Lakes

Crow Wing R.

Ft. Ripley

Mille Lacs

Sauk Rapids

Rum River

Capt. J. Pope 1849

Red River

Wild Rice River

Lieut. C. Grover 1853

Gov. Stevens 1853

White Bear L.

Chippewa R.

Ft. Snelling

St. Paul

St. Croix R.

Spirit Hill

J. N. Nicollet

DAKOTAS

Ft. Ridgely

Minnesota (St. Peters) River

Yellow Medicine R.

ISANTIE DAKOTAS

WINNEBAGOES

44°

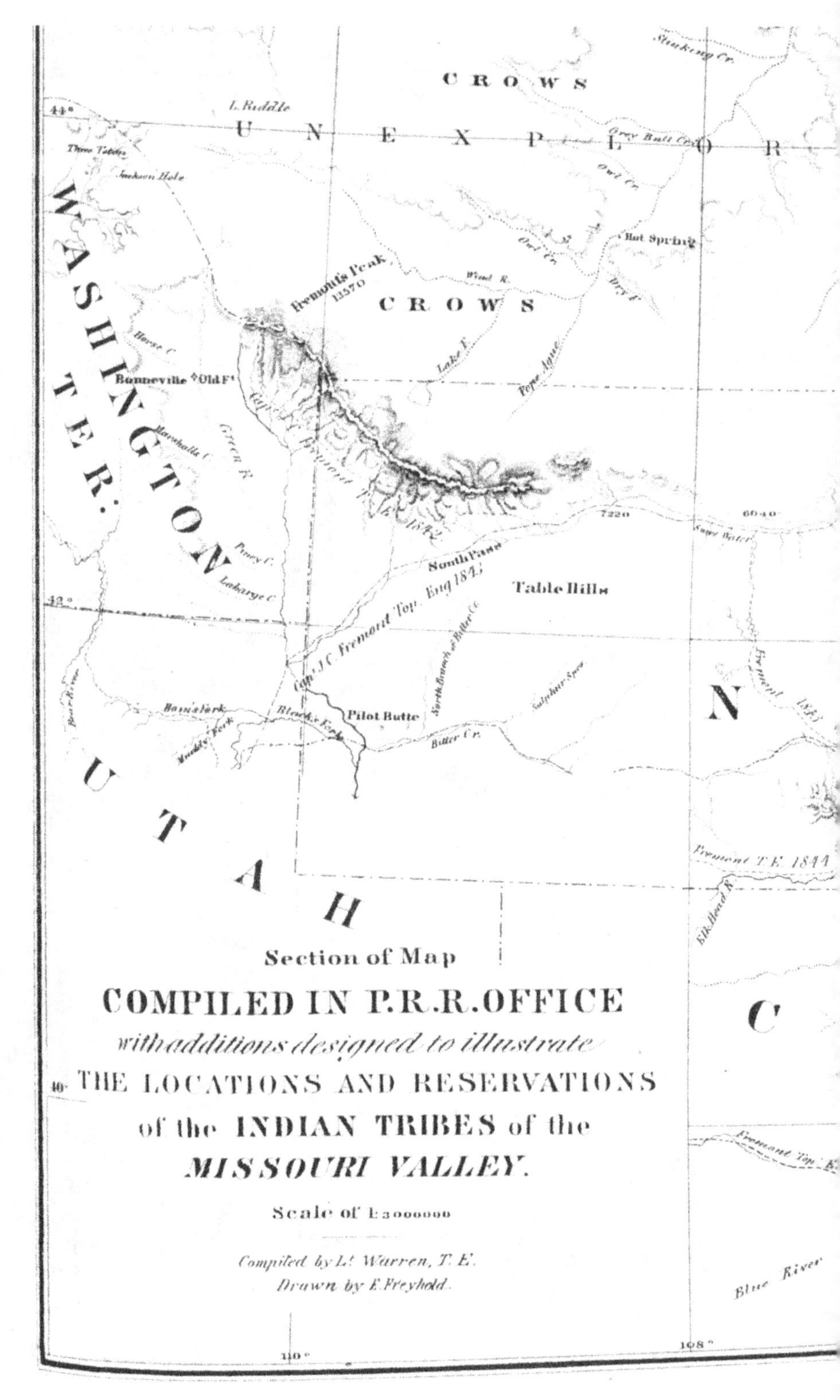
CROWS
UNEXPLOR
CROWS
WASHINGTON TER:
UTAH
N
C
Fremont's Peak 13570
Bonneville Old Ft
Green R.
South Pass
Table Hills
Pilot Butte
Jackson Hole
Three Tetons
Hot Spring
Wind R.
Blue River
Fremont T.E. 1844
Capt. J.C. Fremont Top. Eng 1843
7220
6040
44°
42°
40°
110°
108°
Section of Map
COMPILED IN P.R.R. OFFICE
with additions designed to illustrate
THE LOCATIONS AND RESERVATIONS
of the INDIAN TRIBES of the
MISSOURI VALLEY.
Scale of 1:3000000
Compiled by Lt. Warren, T. E.
Drawn by E. Freyhold.

Powder R.
BLACK
AND
HILLS
SANSARCS
DAKOTAS
Powder R. Butte
Bear Peak
Elk C.
Box Elder C.
Dry F.
Lance C.
Old Woman C.
Sage C.
Butte Cachée
White River
Capt. H. Stansbury Topl. Eng. 1849
Red Buttes
G.K. Warren Topl. Eng. 1855
Dancers Hill
Raw Hide Peak
L'Eau qui court or Rapid R.
Ft. Laramie
1950
Laramie R.
BLACK
HILLS
RAMIE PLAINS
Scotts Bluff
Chimney Rock
North Fork
Ancient Ruins
of Platte
Capt. J.C. Fremont Topl. Eng.
Capt. J.C. Fremont 1842
OGALALA DAKO
Lodge Pole Cr.
Cheyenne Pass
North Fork of Platte
NORTH PARK
MIDDLE
PARK
Long Peak
4030
St. Vrains Ft.
Maj. Long 1820
4000
South Fork of Platte
Republican Fork
SHYENNES
AND
ARAPAHOS
104°

Old Ft George
Lit. Missouri R.
BRULÈ
S I O U
DAKOTAS
White R.
Ft Lookout
2nd Cedar I.
White lakes
Bijou Hill
IHANKTO
DAKOTAS
1st Cedar I.
YANCTONS
OR
UNEXPLORED
Dogs Ears
White L.
2310
Eova Paha
Turtle Hill
Ft Randall
The Tower
PONKAS
Ponka R.
L'Eau qui Court or Rapid R.
Long Pine Cr.
Great
Sand
Hills
G. K. Warren
Pawnee Loup or Wolf R.
A
S
K
PAWNEES
General Harney's Fight with the Indians Sept 3rd 1855
Ft Grattan
TAS
NEBRASKA or PLATTE RIVER
Maj. Long 1820
O'Fallons Bluff
Bradys Island
Grand Island
Ft Kearny
Route of Capt. Fremont 1843
Republican Fork
KANSAS
Solomons Fork
Grand Saline Fork
102°
100°

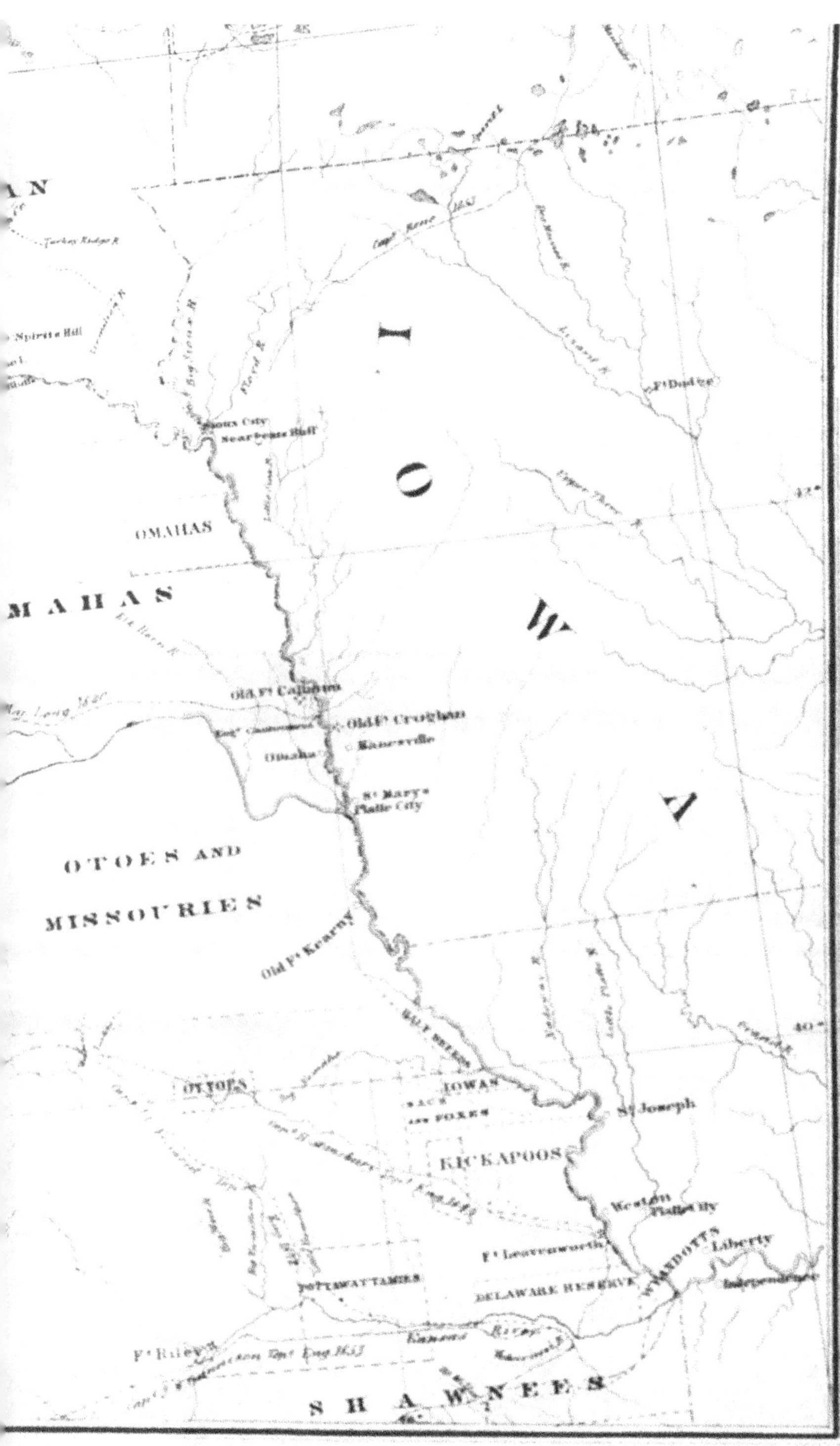

I O W A
OMAHAS
MAHAS
Old Ft Calhoun
Old Ft Croghan
Kanesville
Omaha
St Mary's
Platte City
Sioux City
Ft Dodge
OTOES AND
MISSOURIES
Old Ft Kearny
OTTOES
IOWAS
SACS AND FOXES
St Joseph
KICKAPOOS
Weston
Platte City
Ft Leavenworth
WYANDOTTS
Liberty
Independence
POTTAWATTAMIES
DELAWARE RESERVE
Kansas River
Ft Riley
S H A W N E E S
40°

www.ingramcontent.com/pod-product-compliance
Lightning Source LLC
Chambersburg PA
CBHW081337190326
41458CB00018B/6034
9798893984552